HOMEWORK HELPERS

Trigonometry

WITHDRAWN

DENISE SZECSEI

CAREER
PRESS
Franklin Lakes NJ

HOMEWORK HELPERS: TRIGONOMETRY
EDITED BY KATE HENCHES
TYPESET BY EILEEN DOW MUNSON
Cover design by Lu Rossman/Digi Dog Design
Printed in the U.S.A. by Book-mart Press

To order this title, please call toll-free 1-800-CAREER-1 (NJ and Canada: 201-848-0310) to order using VISA or MasterCard, or for further information on books from Career Press.

The Career Press, Inc., 3 Tice Road, PO Box 687,
Franklin Lakes, NJ 07417
www.careerpress.com

Library of Congress Cataloging-in-Publication Data

Szecsei, Denise.
     Homework helpers : Trigonometry / by Denise Szecsei.
        p. cm.
     Includes index.
     ISBN-13: 978-1-56414-913-8
     ISBN-10: 1-56414-913-7
        1. Trigonometry. 2. Trigonometry—Problems, exercises, etc. I. Title.

QA531.S94 2006
516—dc22
                                                              2006028234

Acknowledgments

Many people were involved in the production of this book, and I would like to thank the people who helped throughout the entire production.

I would like to thank Michael Pye, Kristen Parkes, and everyone else at *Career Press* for helping to turn my manuscript into the book you are holding in your hand. Jessica Faust handled the logistics so that I could focus on writing.

Kendelyn Michaels took time out of her busy schedule to help me polish my writing. Alic Szecsei helped reduce the number of typographical errors in the manuscript and was willing to spend his summer working out the solutions to the review problems.

My family continues to be patient with me, especially as my deadlines approach. Their willingness to take on extra responsibilities while I take care of the last-minute details does not go unnoticed.

Contents

Preface

Welcome to *Homework Helpers: Trigonometry!*

Trigonometry is a hybrid subject. It is based on a mixture of both geometry and algebra, and it involves the study of angles, triangles, and circles.

There are many applications of trigonometry. Global positioning systems use trigonometry to help a person navigate across town. Radar is another application trigonometry: it can be used to track planes and detect the speed of passing motorists. Physicists and engineers use trigonometry to analyze force diagrams and calculate the work done in moving an object. In addition to these practical applications of trigonometry, mathematicians use some of the methods developed in trigonometry to examine a wide variety of abstract mathematical systems including vectors and complex numbers. These applications of trigonometry primarily involve triangles.

Trigonometry can also be used to model repetitive or cyclic behavior. The orbit of a planet around its sun, the number of daylight hours, or the height of the high and low tides of the ocean can all be modeled using trigonometry. One of the strengths of trigonometry is that it is particularly useful when analyzing any type of phenomenon that exhibits circular patterns.

Learning trigonometry involves looking at problems from a new perspective. By combining the concepts in geometry with the problem-solving skills learned in algebra, trigonometry will enable you to solve some very interesting problems. It may take some practice putting these two branches of mathematics together, and this book will serve to guide you through the subject.

I wrote this book with the hope that it will help anyone who is struggling to understand trigonometry or is just curious about the subject. Reading an ordinary math book can be a challenge, so I tried to use everyday language to explain the concepts being discussed. Looking at solutions to math problems is often confusing, so I explain each of the steps I use to get from Point A to Point B. Keep in mind that learning mathematics is not a spectator sport. In this book I have worked out many examples, and I have supplied practice problems at the end of most of the lessons. Work these problems out on your own, and check your answers against the solutions at the end of each chapter. Your answer and my answer should match.

I hope that in reading this book you will develop an appreciation for the subject of trigonometry and the field of mathematics!

Functions

The concept of a function is very important in mathematics. A function is a set of instructions for how to combine numbers. Functions can be used to describe, or model, many situations in our everyday lives. Functions are introduced in algebra, and they are part of almost every aspect of mathematics. There are six new functions that we will define and analyze in trigonometry. These trigonometric functions are useful in modeling behavior that is repetitive, or cyclic. These functions can be used to predict the occurrence of the next lunar eclipse or the number of hours of daylight. They can also be used to describe the sound made when a guitar string is plucked. In order to study these trigonometric functions, we will need a solid understanding of the properties of functions in general.

Lesson 1-1: Representing Functions

A **function** is a set of instructions that establishes a relationship between two quantities. A function has input and output values. The input is called the **domain** and the output is called the **range**. The variable used to describe the elements in the domain is called the **independent variable**, which is commonly represented by x. The variable used to describe the output is called the **dependent variable**, as it *depends* on the input. An important feature of a function is that every input value has only one corresponding output value. A function can be represented in a variety of ways. Functions can be described using words, a formula, a table, or a graph.

Using a formula to define a function is a convenient way to describe the function in mathematical terms instead of using words. When analyzing a formula, it is important to use the order of operations. A function is often written $y = f(x)$, where $f(x)$ is an expression that involves the independent variable x. For example, the formula $y = 2x + 1$ defines a linear function. The name of this function is y. We could also have written this function as $f(x) = 2x + 1$. In this case, the name of the function is $f(x)$. We use the notation y and $f(x)$ interchangeably. When we identify a function using the notation $f(x)$, the variable in parentheses, or the independent variable (in this case, x), is sometimes referred to as the **argument** of the function. Evaluating a function for a particular value of x involves replacing every instance of the independent variable with that value.

Scientists often collect data from various instruments and record this information in a table. These tables represent functions and provide an easy way to describe a complex, or unknown formula.

The graph of a function is usually presented using the Cartesian coordinate system. In the Cartesian coordinate system, we use a vertical line, called the **y-axis**, and a horizontal line, called the **x-axis**, to divide the plane into four regions, or **quadrants**. The intersection of these two lines is called the **origin**.

Two numbers are used to describe the location of a point in the plane, and they are recorded in the form of an ordered pair (x, y). A function can then be thought of as a collection of ordered pairs (x, y). The graph of a function is then the graph of these ordered pairs on the coordinate plane.

If the graph of the function $f(x)$ crosses the x-axis, then we say that $f(x)$ has an x-intercept. The **x-intercept** of a function is the point where the function crosses the x-axis. Because the y-coordinate of any point on the x-axis is 0, the x-intercept of a function corresponds to a point of the form $(a, 0)$. The x-intercepts of a function are often called the **roots** of the function, or the **zeros** of the function. Not all functions have x-intercepts. Finding the x-intercepts of a function involves solving the equation $f(x) = 0$ for x.

If $x = 0$ is in the domain of a function $f(x)$, then the point $(0, f(0))$ is the y-intercept of $f(x)$. The **y-intercept** of a function represents where the function crosses the y-axis. Not all functions have y-intercepts, but if a function has a y-intercept, the y-intercept will be unique. In other words, a function can have, at most, one y-intercept.

The graph of an equation does not necessarily have to be a *function*. For example, the graph of the equation $x^2 + y^2 = 1$ is a circle of radius 1 centered at the origin. The graph of this equation is not a function because it fails the **vertical line test**. If any vertical line intersects a graph at more than one point, then the graph is not the graph of a function. The vertical line $x = 0$ intersects the graph of the equation $x^2 + y^2 = 1$ in two places: $y = 1$ and $y = -1$. The vertical line test is a physical interpretation of the fact that, with a function, every input value, x, has only one corresponding output value, y.

Lesson 1-2: The Domain of a Function

The **domain** of a function represents the allowed values of the independent variable, x. If a function is described using words, then the domain needs to incorporate the context of the description of the function. For example, if a function describes the amount of electricity needed to be generated as a function of the number of houses on the power grid, then the domain of this function cannot include any negative numbers: providing electricity to a negative number of houses makes no sense!

The description of a function using a formula may or may not include a domain. If the domain is not indicated, then it is safe to assume that the domain is the set of all real numbers that, when substituted in for the independent variable, produce real values for the dependent variable. In general, to find the domain of a function, start with the set of all real numbers and whittle down the list. There are two things that are frowned upon in the mathematical community. The first thing that is forbidden is to divide a non-zero number by 0; quotients such as $\frac{3}{0}$ are meaningless. The second thing that is not allowed in the world of real numbers is to take an even root of a negative number; for example, there is no real number that corresponds to $\sqrt{-4}$. To find the domain of a function that involves an even root (a square root, a fourth root, and so on) set whatever is under the radical to be greater than or equal to 0 and find the solutions to the inequality. Then toss out any points that result in the denominator being equal to 0. We will make use of these ideas when we analyze the domain of the trigonometric functions.

If the domain of a function is to be limited, it is important to write it down in a way that can be easily understood by others. One common way to describe the domain of a function uses interval notation. The following table summarizes the different types of intervals that we will encounter. Parentheses mean that everything right up to that point is included. Brackets mean everything right up to, and including the point, is included. Keep in mind that parentheses are always used next to the symbol for infinity, ∞. Brackets are never used next to ∞ because x can never actually reach infinity.

Interval	Name	Meaning	Examples
(a,b)	Open	All values of x that satisfy the inequality $a < x < b$ are contained in the interval. The endpoints <u>are not</u> contained in the interval.	$(1,2), (-\infty, \infty),$ $(-\infty, 2), (3, \infty)$
$[a,b]$	Closed	All values of x that satisfy the inequality $a \leq x \leq b$ are contained in the interval. The endpoints <u>are</u> contained in the interval.	$[-2, 5]$
$(a,b]$	Half-open (or half-closed)	All values of x that satisfy the inequality $a < x \leq b$ are contained in the interval. Only the <u>right</u> endpoint is contained in the interval.	$(-\infty, 3], (2, 9]$
$[a,b)$	Half-open (or half-closed)	All values of x that satisfy the inequality $a \leq x < b$ are contained in the interval. Only the <u>left</u> endpoint is contained in the interval.	$[-3, \infty), [3, 7)$

Lesson 1-3: Operations on Functions

The *algebra* of real numbers establishes rules for how to combine real numbers. Numbers can be added, subtracted, multiplied, and divided. Addition, subtraction, multiplication, and division are examples of some of the common **operations** that we perform on real numbers. Algebraic expressions are an abstract way to represent numbers, so it is only natural that we are able to add, subtract, multiply, and divide algebraic expressions as well. There is one additional thing that we can do with functions: we can take their *composition*.

A function is a transformation of things from the domain, or the input, to things in the range, or the output. If a function f has a domain X and its range is a subset of a set Y, then we use the notation $f: X \rightarrow Y$ to represent the idea that f is a function from X to Y. Suppose that g is a function whose domain is Y and whose range is a subset of a set Z. We can use the functions f and g to define a new function whose domain is X and whose range is contained in Z. This new function would take an element in X to its corresponding element in Y using the function f and then take that element in Y to an element in Z using the function g. This process of stringing functions from set to set is called the **composition** of functions. Because f takes things from X to Y, and g takes things from Y to Z, then the new function "g composed with f" takes things *directly* from X to Z.

We write the composition of f and g in the order described previously as $g \circ f$. The functions are applied *right* to *left*: $g \circ f(x)$ means *first* apply the function f to x, and *then* apply the function g to the result. We can write $g \circ f(x) = g(f(x))$; $g \circ f(x)$ is read "g composed with f," or as "g of $f(x)$." The order in which we compose things matters. In general, $g \circ f(x) \neq f \circ g(x)$. In other words, $g(f(x)) \neq f(g(x))$.

We can look at a complicated function such as $h(x) = \sqrt{3x+1}$ in terms of the composition of two functions. If we define $f(x) = 3x + 1$ and $g(x) = \sqrt{x}$, then $h = g \circ f$. Functions are just instructions for what to do with the argument, or the object in parentheses. The function $g(x) = \sqrt{x}$ instructs us to take the argument of the function and put it under a radical. The function $f(x) = 3x + 1$ instructs us to triple the argument and then add 1. So, if we first apply g to f, and then apply f to x, the result will be h:

$$h(x) = g(f(x)) = \sqrt{f(x)} = \sqrt{3x+1} .$$

Alternatively, we could first apply f to x, and then apply g to $f(x)$. Essentially, this means that we will first substitute in for $f(x)$ using its formula and then apply g:

$$h(x) = g(f(x)) = g(3x+1) = \sqrt{3x+1} .$$

Either method that you use to evaluate the composition $g \circ f(x)$ will result in the function $h(x)$.

Now let's look at the composition in reverse order. Notice that, in this situation, if we first apply f to $g(x)$ and then substitute in for $g(x)$, we have:

$$f \circ g(x) = f(g(x)) = 3g(x) + 1 = 3\sqrt{x} + 1.$$

Alternatively, if we first substitute in for $g(x)$ using its formula and then apply f, we have:

$$f \circ g(x) = f(g(x)) = f(\sqrt{x}) = 3\sqrt{x} + 1.$$

This illustrates the fact that the order in which you *compose* functions matters, but the order that you *substitute* in for f and g does not. In the example we just looked at,

$$g \circ f(x) = \sqrt{3x + 1},$$

whereas $f \circ g(x) = 3\sqrt{x} + 1.$

In general, $f \circ g(x)$ is a different function than $g \circ f(x).$

Lesson 1-3 Review

Find $g \circ f(x)$ and $f \circ g(x)$ for the following pairs of functions:

1. $f(x) = \sqrt{6 - x}$, $g(x) = 2x + 1$

2. $f(x) = \dfrac{1}{3x + 6}$, $g(x) = \frac{1}{3}(x - 6)$

3. $f(x) = x^2 + 2$, $g(x) = \sqrt{x - 2}$

Lesson 1-4: Transformations of Functions

The coordinate plane is a convenient way to organize the points in a plane. It is important to realize that the coordinate plane serves as a reference for locating the points in the plane. The main points of reference are the coordinate axes and the origin. There are times when the formulas used to describe a function are difficult to understand because of how the graph is oriented with respect to the coordinate axes. In Chapter 9, we will learn how to re-orient the coordinate axes to improve our graphing capabilities. To help develop our graphing skills, it is worthwhile to learn how to move the graph of a function around the coordinate plane. The process by which a graph is moved is referred to as a **transformation** of a function. In general, the transformation of the graph of a function can involve a **shift** (also called a translation), a **reflection**, a **stretch,** or a **contraction.** Transformations can occur vertically, meaning that the result is a change in the value of the y-coordinate, or horizontally, meaning that they result in a change in the x-coordinate of the graph. We will look at each of these transformations individually.

To shift a function *vertically*, add a constant to the dependent variable, or to the function: the graph of $f(x) + c$ is the graph of $f(x)$ shifted up c units (if $c > 0$), or down c units (if $c < 0$). To shift a function *horizontally*, add a constant to the *argument* of the function: the graph of $f(x + c)$ is the graph of $f(x)$ shifted to the left c units (if $c > 0$), or to the right c units (if $c < 0$).

Reflecting a graph is like looking at it in a mirror. Every point on one side of the mirror is sent to the corresponding point on the opposite side of the mirror. To reflect the graph of a function across the x-axis, multiply the function by -1: the graph of $-f(x)$ is the graph of $f(x)$ reflected across the x-axis. To reflect the graph of a function across the y-axis, the signs of the x-coordinates of every point of the function must be changed, while the y-coordinates of the points remain the same. Changing the sign of the x-coordinate of a point is equivalent to changing the sign of the independent variable in the formula for the function. In other words, when reflecting a graph across the y-axis, the sign of the *argument* of the function must change. The graph of the function $f(-x)$ is the graph of $f(x)$ reflected across the y-axis.

The effect of *stretching* a graph is to "draw it out" so that it has the same general shape, but occupies more space. *Contracting* a graph effectively shrinks the graph, or makes it narrower, while retaining its general shape. We can stretch or contract a graph vertically or horizontally. To *stretch* the graph of a function *vertically*, multiply the function by a constant that is greater than 1. To *contract* a function vertically, multiply the function by a positive constant that is less than 1. To stretch the graph of a function *horizontally*, multiply the *argument* of the function by a positive constant that is less than 1. To contract the graph of a function *horizontally*, multiply the argument of the function by a constant that is greater than 1. This information is summarized in the following table. We will discuss the various transformations of the graph of the function $f(x) = x^2 + x$.

Effect	Method	Example	Formula
Vertical shift	Add a constant c to the function	Shift the graph up 2 units	$f(x)+2=x^2+x+2$
Horizontal shift	Add a constant c to the argument of the function	Shift the graph to the right 3 units	$f(x-3)=(x-3)^2+(x-3)$
Reflect across the x-axis	Multiply the function by -1	Reflect across the x-axis	$-f(x)=-(x^2+x)$
Reflect across the y-axis	Multiply the argument of the function by -1	Reflect across the y-axis	$f(-x)=(-x)^2+(-x)$
Vertical stretch	Multiply the function by a constant that is greater than 1	Stretch the function vertically by a factor of 4	$4f(x)=4(x^2+x)$
Vertical shrink	Multiply the function by a positive constant that is less than 1	Shrink the function vertically by a factor of 4	$\frac{1}{4}f(x)=\frac{1}{4}\left(x^2+x\right)$
Horizontal stretch	Multiply the argument of the function by a positive constant that is less than 1	Stretch the function horizontally by a factor of 4	$f\left(\frac{1}{4}x\right)=\left(\frac{1}{4}x\right)^2+\left(\frac{1}{4}x\right)$
Horizontal shrink	Multiply the argument of the function by a constant that is greater than 1	Shrink the function horizontally by a factor of 4	$f(4x)=(4x)^2+(4x)$

The transformation of a function can be thought of in terms of composition. Shifting the graph of a function $f(x)$ up 3 units can be interpreted as $g \circ f(x)$, where $g(x) = x + 3$: $g \circ f(x)=g(f(x))=f(x)+3$. Reflecting the graph of a function $f(x)$ with respect to the y-axis can be interpreted as $f \circ g(x)$, where $g(x) = -x$: $f \circ g(x)=f(g(x))=f(-x)$.

Stretching and contracting functions can be interpreted similarly. It is helpful to have several ways to think of the transformation process. Sometimes one perspective works better than another.

Lesson 1-4 Review

Transform the following functions:

1. Shift $f(x) = x^3$ to the right 4 units.

2. Shift $f(x) = 3x - 1$ up 8 units.

3. Shift $f(x) = \sqrt{x}$ to the left 2 units.

4. Shift $f(x) = \frac{1}{x}$ down 5 units.

Lesson 1-5: Symmetry

The graph of a function can have some inherent symmetry. There are two important symmetries that we will discuss in this lesson. The first type of symmetry is when the *y*-axis serves as a mirror. The second type of symmetry has to do with the arrangement of the points of a function relative to the origin.

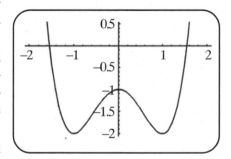

In the graph of the function shown in Figure 1.1, the *y*-axis acts as a mirror. Every point to the left of the *y*-axis has a corresponding point to the right.

The *x*-coordinates of the point on the left and its corresponding point on *Figure 1.1.* the right have the same magnitude; only their signs are different. The *y*-coordinates of the point on the left and its corresponding point on the right are exactly the same. Reflecting the graph of this function across the *y*-axis will not change the function. Recall that the transformation of reflecting across the *y*-axis can be written as $f(-x)$. Symmetry with respect to reflection across the *y*-axis can be characterized algebraically as $f(-x) = f(x)$. If a function is symmetric with respect to the y-axis, that function is called an **even** function.

The graph of the function shown in Figure 1.2 on page 20 is symmetric with respect to the origin. Notice that points in Quadrant I have

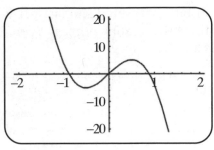

Figure 1.2.

corresponding points in Quadrant III, and points in Quadrant II have corresponding points in Quadrant IV. This can be seen as reflecting a point across *both* the *x*-axis and the *y*-axis.

Reflecting across the *y*-axis can be written as $f(-x)$, and reflecting across the *x*-axis can be written as $-f(x)$. Combining these transformations results in the overall transformation $-f(-x)$. The only point that does not change when you reflect it across both axes is the origin, which is why we refer to this type of symmetry as being symmetric with respect to the origin. A function that is symmetric with respect to the origin must satisfy the equation $f(x) = -f(-x)$. This is often rewritten as $f(-x) = -f(x)$. Functions that are symmetric with respect to the origin are called **odd** functions.

Graphing a function that is symmetric requires less work if you use the symmetry to your advantage. You can plot half as many points and still see the whole picture. To graph an even function, find the y-intercept and then evaluate the function for some positive values of x. Use the symmetry of the function to graph the corresponding points to the left of the *y*-axis.

To graph an odd function, first realize that every odd function must pass through the origin. This can be seen by evaluating the equation $f(-x) = -f(x)$ when $x = 0$:

$$f(-x) = -f(x)$$
$$f(-0) = -f(0)$$
$$f(0) = -f(0)$$

The only number that is unchanged when you multiply by -1 is 0, so $f(0) = 0$ and we see that an odd function must pass through the origin. To graph an odd function, evaluate the function for some positive values of x. Use the symmetry of the function and the fact that the graph of the function must pass through the origin to graph the corresponding points.

Lesson 1-6: Properties of Functions

When analyzing a function, it is important to know where the graph of the function is rising and where it is falling. A function is *increasing* if the value of the function increases as the value of the independent variable increase. A function is increasing if, as you look at the graph from left to right, the function gets higher. We can state this definition more precisely:

A function $f(t)$ is **increasing** on an interval I if
$f(a) < f(b)$ whenever $a < b$.

Similarly, a function is *decreasing* if the value of the function decreases as the value of the independent variable increase. A function is decreasing if, as you look at the graph from left to right, the function gets lower. This can also be stated precisely:

A function $f(t)$ is **decreasing** on an interval I if
$f(a) > f(b)$ whenever $a < b$.

A function is constant on an interval if the value of the function does not change on the interval.

A function $f(t)$ is **constant** on an interval I if
$f(a) = f(b)$ for all a and b in I.

A function is **monotonic** on an interval if it is either always increasing or always decreasing on that interval. In other words, a monotonic function does not change directions. Linear functions are monotonic on their entire domain. A linear function will be monotone *increasing* if it has a positive slope, and it will be monotone *decreasing* if it has a negative slope.

Fortunately, the definitions for increasing and decreasing functions do not involve advanced mathematics. A comparison of function values at various points in the domain will determine whether or not a function is increasing, decreasing, or neither, on a particular interval I. Unfortunately, evaluating a function at a variety of points in the domain can become tedious.

When the graph of a function makes the transition from increasing to decreasing, the function will reach a maximum at the point the transition occurs. This point is called a *local maximum*. Conceptually, a function has a local maximum at a point $x = c$ if the function gets no bigger than $f(c)$ on an open interval that contains c. A local maximum looks like the top of a little hill. Figure 1.3 shows the graph of a function with a local maximum.

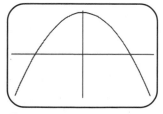

Figure 1.3.

> A function $f(t)$ has a **local maximum** at $x = c$ if there is an open interval I containing c so that, for all t in I, $f(t) \leq f(c)$.

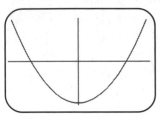

Figure 1.4.

When the graph of a function makes the transition from decreasing to increasing, the function will reach a minimum at the point at which the transition occurs. This point is called a *local minimum*. Conceptually, a function has a local minimum at a point $x = c$ if the function gets no lower than $f(c)$ on an open interval that contains c. A local minimum looks like a little valley. Figure 1.4 shows the graph of a function with a local minimum.

> A function $f(t)$ has a **local minimum** at $x = c$ if there is an open interval I containing c so that, for all t in I, $f(c) \leq f(t)$.

The graph of a function can have several local maxima and minima, as is the case with the graph shown in Figure 1.5. This function has local maxima at $x = -2$ and $x = 4$, and local minima at $x = -4$ and $x = 2$.

Shifting a function vertically will change the *values* of the local maxima and minima, but will not change their *location*. If a function has a local

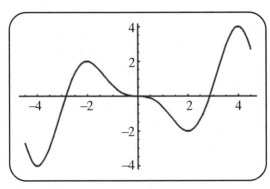

Figure 1.5

maximum or minimum at the point (a, b), and its graph is shifted vertically c units, then the local maximum or minimum will be the point $(a, b + c)$. Shifting a function horizontally will change the *location* of the local maxima and minima, but will not change their *values*. If a function has a local maximum or minimum at the point (a, b), and its graph is shifted horizontally c units, then the local maximum or minimum will be the point $(a - c, b)$. Reflecting a graph with respect to the x-axis will turn local maxima into local minima, and local minima into local maxima. Reflecting a graph with respect to the y-axis will change the *location* of the local maxima and minima (meaning the x-coordinate of their location), but will not change their *values* (meaning the y-coordinate of their location).

Another important property of a function is whether or not it is one-to-one. A function is **one-to-one** if any two different values in the domain correspond to different values in the range. This can also be thought of as one input value resulting in a given output value. The function $f(x) = x^2$ is not one-to-one because the values $x = 2$ and $x = -2$ correspond to the same value in the range $f(2) = 4$ and $f(-2) = 4$. Functions that are one-to-one will have an inverse. Functions that are not one-to-one may have an inverse if the domain is restricted to a region where the function is one-to-one. The inverse of a function that is not one-to-one will depend on the restriction of the domain. For example, if the domain of the function $f(x) = x^2$ is restricted to the non-negative real numbers, then its inverse will be $\sqrt{x}$. If the domain is restricted to the non-positive real numbers, then its inverse will be $-\sqrt{x}$.

The horizontal line test can be used to determine whether or not a function is one-to-one. If a horizontal line intersects the graph of a function at more than one point, then the function is not one-to-one. There is nothing wrong if a horizontal line does not intersect a function at all. The only issue is whether or not a horizontal line intersects the graph of a function at two or more places. If so, then the function will not be one-to-one, and hence will not be invertible on that domain. We will apply these ideas to the trigonometric functions in Chapter 4.

Answer Key
Lesson 1-3 Review

1. $g \circ f(x) = g(f(x)) = 2f(x) + 1 = 2\sqrt{6 - 1} + 1,$

$f \circ g(x) = f(g(x)) = \sqrt{6 - g(x)} = \sqrt{6 - (2x + 1)} = \sqrt{5 - 2x}$

2. $g \circ f(x) = g(f(x)) = \frac{1}{3}\left(\frac{1}{3x+6} - 6\right) = \frac{1}{3}\left(\frac{1}{3x+6} - \frac{6(3x+6)}{(3x+6)}\right) = \frac{1}{3}\left(\frac{-18x-35}{3x+6}\right)$

$f \circ g(x) = f(g(x)) = \frac{1}{3g(x)+6} = \frac{1}{3\left(\frac{1}{3}(x-6)\right)+6} = \frac{1}{(x-6)+6} = \frac{1}{x}$

3. $g \circ f(x) = g(f(x)) = \sqrt{f(x) - 2} = \sqrt{(x^2 + 2) - 2} = \sqrt{x^2} = x$

$$f \circ g(x) = f(g(x)) = [g(x)]^2 + 2 = \left(\sqrt{x-2}\right)^2 + 2 = (x-2) + 2 = x$$

Lesson 1-4 Review

1. $f(x-4) = (x-4)^3$

2. $f(x) + 8 = (3x-1) + 8 = 3x + 7$

3. $f(x+2) = \sqrt{x+2}$

4. $f(x) - 5 = \frac{1}{x} - 5$

Right Triangle Trigonometry

Geometry plays an important role in trigonometry. Many of the results in trigonometry are based on the properties of right triangles and circles. In fact, the two most common approaches to trigonometry are referred to as the *right triangle* approach and the *unit circle* approach.

Trigonometry brings together the best of algebra and geometry. We will begin this chapter by reviewing some analytical results from geometry, including the Pythagorean Theorem. These basic ideas will enable us to develop trigonometry using the right triangle approach.

Lesson 2-1: Angles and Their Measures

A ray is half of a line. A ray begins at a point and continues forever in one direction. The starting point of the ray is called the **vertex**, or **endpoint**, of the ray. If two rays share a common vertex, they form an angle. One of the rays is called the **initial side** of the angle, and the other ray is called the **terminal side** of the angle. The angle formed is identified by indicating the direction and the amount of rotation from the initial side to the terminal side. By convention, if the rotation is in the *counterclockwise* direction, the measure of the angle is *positive*. If the rotation is *clockwise*, then the measure of the angle is *negative*.

Geometers tend to focus on the interior angle formed, and the measure of an angle is usually a *positive* number between 0 and 180. You may recall that angles can be classified using the terms *acute*, *right*, and *obtuse*. An **acute** angle is an angle whose measure is between 0° and 90°, a **right** angle is an angle whose measure is exactly 90°, and an **obtuse** angle is an angle whose measure is between 90° and 180°. An angle of measure 0° is meaningless in geometry, and if an angle has measure 180°, then the initial side and the terminal side are **collinear**, meaning that they lie on the same line. Two angles are **complementary** if their sum is a right angle. In other words, angles α and β are complementary if $\alpha + \beta = 90°$.

In trigonometry, our angle measurements will not be so limited. An angle with measure 360° corresponds to an angle whose initial side and terminal side coincide, and whose rotation is in the counterclockwise direction. An angle with measure −360° corresponds to an angle whose initial side and terminal side also coincide, but its rotation is in the clockwise direction. An angle with measure 0° corresponds to an angle whose initial side and terminal side coincide, and there is no rotation. All three of these angles are shown in Figure 2.1.

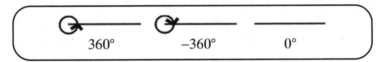

$$360° \qquad -360° \qquad 0°$$

Figure 2.1.

We typically use lowercase Greek letters to indicate an angle. The most common Greek letters used to denote an angle are α (alpha), β (beta), γ (gamma), δ (delta), θ (theta), and ϕ (phi). Figure 2.2 shows three angles: α, θ, and ϕ. The angle α has positive measure and is the interior angle of the two rays that form the angle. The angle θ also has positive measure, but it is called the reflex angle of the two rays that form the angle. The angle ϕ has negative measure.

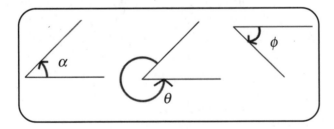

Figure 2.2.

The Cartesian coordinate system is instrumental in bringing together algebra and geometry. The Cartesian coordinate system enables us to orient geometric shapes and measure lengths and angles. An angle is in standard position if its vertex is located at the origin of the Cartesian coordinate system and its initial side coincides with the positive x-axis. Figure 2.3 shows some angles in standard position.

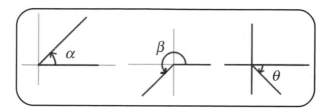

Figure 2.3.

When an angle is in standard position, its terminal side will lie in one of the four quadrants of the plane. An angle lies in the quadrant that its terminal side lies in, when it is standard position. In Figure 2.3, α lies in Quadrant I, β lies in Quadrant III, and θ lies in Quadrant IV. If the terminal side of an angle lies on either the *x*-axis or the *y*-axis, it is called a **quadrantal angle.**

The measure of an angle is determined by measuring the amount of rotation needed for the initial side to coincide with the terminal side. The two most commonly used measures for angles are degrees and radians. Both of these units are defined in terms of the circumference of a circle.

> An angle of 1 **degree**, written 1°, is defined to be the angle at the center of a circle that cuts off an arc having a length of $\frac{1}{360}$ of the circumference of the circle.

In other words, the degree is a unit of measurement that represents $\frac{1}{360}$ of one full rotation around a circle. Nautical and astronomical applications of trigonometry are two areas where the measure of an angle is given in units of degrees, minutes, and seconds.

The other useful unit of measurement for an angle is the radian.

> One **radian** is defined to be angle at the center of a circle that cuts off an arc having a length equal to the radius of the circle.

Radians are more useful in some applications of trigonometry because the formulas derived take on a much cleaner form. For example, radians are particularly useful in calculating the arc length and the area of a sector. If a circle has a radius, *r*, and the arc cuts off an angle θ, as shown in Figure 2.4 on page 28, then the arc length, *s*, can be found by the formula $s = r\theta$, and the area of the sector, *A*, can be found by the formula $A = \frac{1}{2}r^2\theta$.

Another application of trigonometry where radians are preferable to degrees is in calculating the linear speed of an object traveling in circular motion. The **average speed** of an object is the distance traveled divided by the elapsed time. If an object moves at a constant speed along a circular path, then the **linear** speed of the object, v, is defined as the arc length traveled, s, divided by the time elapsed,

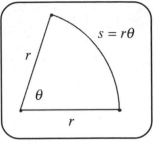

Figure 2.4.

t: $v = \frac{s}{t}$. The distance traveled is the arc length of the circle. The arc length traveled is given by the formula $s = r\theta$, where θ is the measure of the central angle of the circular path traced out by the object. If we combine these two equations, we have:

$$v = \frac{s}{t} = \frac{r\theta}{t} = r\left(\frac{\theta}{t}\right)$$

The quantity $\frac{\theta}{t}$ is called the **angular** speed of the object, and is denoted by the Greek letter ω (omega): $\omega = \frac{\theta}{t}$. Angular speed is the rate at which the central angle θ is changing. Angular speed is the number of radians that the central angle changes divided by the elapsed time. Angular speed and linear speed are related by the formula $v = r\omega$. The units for linear speed are *length per unit time*. The units for angular speed are *radians per unit time*. Angular speed is used to measure cyclic motion, and the units are commonly given as revolutions per minute, or rpm. To convert from revolutions per minute to angular speed, use the fact that one revolution is equivalent to 2π radians.

Example 1

A wheel with diameter equal to 10 cm is turning at a rate of 150 rpm. What is the angular speed of the wheel?

Solution: Convert revolutions per minute to radians per minute:

$$\omega = \frac{150 \text{ revolutions}}{\text{minute}} \cdot \frac{2\pi \text{ radians}}{\text{revolution}} = \frac{300\pi \text{ radians}}{\text{minute}}$$

The angular speed is 300π radians per minute.

Example 2

A Ferris wheel has a radius of 30 feet. If it takes 75 seconds for the wheel to make one revolution, find the linear speed of the people riding the Ferris wheel, in feet per second.

Solution: First, find the angular speed. Then use the equation $v = r\omega$ to find the linear speed:

$$\omega = \frac{1 \text{ revolution}}{75 \text{ seconds}} \cdot \frac{2\pi \text{ radians}}{\text{revolution}} = \frac{2\pi \text{ radians}}{75 \text{ seconds}}$$

$$v = r\omega = (30 \text{ feet}) \cdot \frac{2\pi \text{ radians}}{75 \text{ seconds}} = \frac{4\pi \text{ feet}}{5 \text{ seconds}}$$

The linear speed will be $\dfrac{0.8\pi \text{ feet}}{\text{second}}$, or approximately $2.51 \text{ ft}/\text{sec}$.

Example 3

An auto mechanic is balancing the tires on a car. If the tires have a diameter of 30 inches, and they are being rotated at a rate of 440 rpm, find the linear speed of the tires in miles per hour.

Solution: First, find the angular speed. Then use the equation $v = r\omega$ to find the linear speed:

$$\omega = \frac{440 \text{ revolution}}{\text{minute}} \cdot \frac{2\pi \text{ radians}}{\text{revolution}} = \frac{880\pi \text{ radians}}{\text{minute}}$$

$$v = r\omega = (30 \text{ inches}) \cdot \frac{880\pi \text{ radians}}{\text{minute}} = \frac{26400\pi \text{ inches}}{\text{minute}}$$

We need to convert inches per minute to miles per hour:

$$\frac{26400\pi \text{ inches}}{\text{minute}} \cdot \frac{1 \text{ foot}}{12 \text{ inches}} \cdot \frac{1 \text{ mile}}{5280 \text{ feet}} \cdot \frac{60 \text{ minutes}}{\text{hour}} = \frac{25\pi \text{ miles}}{\text{hour}}$$

The linear speed will be $\dfrac{25\pi \text{ miles}}{\text{hour}}$, or approximately 78.5 mph.

There are times when it is necessary to convert between radians and degrees. To convert between the two units, relate them through the unit

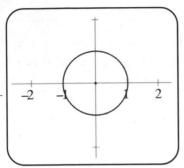

Figure 2.5.

circle. A **unit circle** is a circle that is centered at the origin and has a radius equal to 1 unit, as shown in Figure 2.5.

When measured in degrees, the central angle of a full circle is 360°. When measured in radians, the central angle of a full circle is just the circumference of the circle: because we are working with the unit circle, the circumference is 2π. These two measurements for the central angle of a full circle must agree: $360° = 2\pi$. This means that π radians must correspond to 180°. To convert an angle measurement from radians to degrees, set up a proportion:

$$\frac{\theta_{rad}}{\theta_{deg}} = \frac{\pi}{180}.$$

You can use this proportion to solve for either θ_{rad} or θ_{deg}, depending on which one you are given and which one you need to find.

Example 4

A water sprinkler sprays water over a distance of 25 feet while rotating through an angle of 125°. Find the area of the lawn that receives water from this sprinkler.

Solution: Convert the angle of rotation of the sprinkler from degrees to radians:

$$\frac{\theta_{rad}}{\theta_{deg}} = \frac{\pi}{180}$$

$$\frac{\theta_{rad}}{125} = \frac{\pi}{180}$$

$$\theta_{rad} = \frac{125\pi}{180}$$

$$\theta_{rad} = \frac{25}{36}\pi$$

Next, use the formula for the area of a sector of a circle:

$$A = \frac{1}{2}r^2\theta$$

$$A = \frac{1}{2}(25)^2 \left(\frac{25}{36}\pi \right) = \frac{15625}{72}\pi$$

The area of the lawn is $\frac{15625}{72}\pi$ ft^2, which is approximately 682 ft^2.

Lesson 2-1 Review

1. A child is spinning a rock at the end of a 2-foot rope at the rate of 120 revolutions per minute. Find the linear speed of the rock when it is released.

2. Find the arc length and the area of a sector of a circle of diameter 10 cm and central angle 100°.

3. Find the central angle of a sector of a circle if its arc length is 2π cm and its area measures 10π cm^2.

Lesson 2-2: The Pythagorean Theorem

The Pythagorean Theorem is an important theorem in geometry, and it applies to right triangles. The Pythagorean Theorem states that the square of the length of the hypotenuse of a right triangle is equal to the sum of the squares of the lengths of the two legs. Consider the right triangle shown in Figure 2.6. Suppose that the lengths of the two legs of the right triangle are a and b, respectively, and that the length of the hypotenuse is c. By the Pythagorean theorem, we have the relationship $a^2 + b^2 = c^2$.

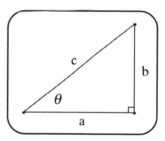

Figure 2.6.

The Pythagorean Theorem can be used to establish a formula to calculate the distance between two points in the coordinate plane. The distance, d, between the points (x_1, y_1) and (x_2, y_2) is given by the formula:

$$d = \sqrt{(x_2 - x_1)^2 + (y_2 - y_1)^2}$$

From the distance formula, we can write a formula to represent all of the points that are a fixed distance, r, from a given point (h, k):

$$\left\{ (x,y) : \sqrt{(x-h)^2 + (y-k)^2} = r \right\}$$

This collection of points is called a **circle**. The point (h, k) is called the **center** of the circle, and the fixed distance, r, is called the **radius** of the circle. We can simplify the equation that appears in the brackets. Squaring both sides of the equation gives the formula:

$$(x - h)^2 + (y - k)^2 = r^2$$

This formula is called the **standard form of a circle**. In this form, the center of the circle and its radius can be determined quickly. We have already briefly discussed one special circle, called the unit circle. It is a circle that is centered at the origin, and has radius equal to 1. Substituting the coordinates of the origin, $(0, 0)$, for the center of the circle, we can establish the equation of the unit circle:

$$x^2 + y^2 = 1$$

Keep in mind that trigonometry is based on the properties of circles, and this equation will play an important role in establishing trigonometric identities.

Lesson 2-3: Trigonometric Ratios of Acute Angles

In the triangle approach to trigonometry, we start with a right triangle, as shown in Figure 2.6 on page 31.

We will define the sine and cosine ratios for acute angles first, and then generalize the results. Suppose that the lengths of the two legs of the right triangle are a and b, respectively, and that the length of the hypotenuse is c. By the Pythagorean Theorem, we have the relationship $a^2 + b^2 = c^2$. If the side with length b is *opposite* the angle θ in the triangle, and the side with length a is *adjacent* to the angle θ, then we

define $\sin\theta = \frac{b}{c}$ and $\cos\theta = \frac{a}{c}$. Right triangles are not always oriented in this way. It is beneficial to understand how to calculate the sine and cosine of an angle for any triangle orientation. Conceptually:

$$\sin\theta = \frac{\text{opposite}}{\text{hypotenuse}}, \text{ and } \cos\theta = \frac{\text{adjacent}}{\text{hypotenuse}}$$

From the Pythagorean Theorem, we see that $a \leq c$ and $b \leq c$, which means that $\sin\theta \leq 1$ and $\cos\theta \leq 1$.

The sine and cosine of certain angles can be found using triangle geometry. For example, we can calculate the sine and cosine of the angles of the right triangles shown in Figure 2.7 on page 33.

Example 1

Calculate the sine and cosine of the angles θ, δ, α, and β as shown in Figure 2.7.

Solution: The sine of an angle is the ratio $\sin\theta = \dfrac{\text{opposite}}{\text{hypotenuse}}$, and the cosine of an angle is the ratio $\cos\theta = \dfrac{\text{adjacent}}{\text{hypotenuse}}$. We will need

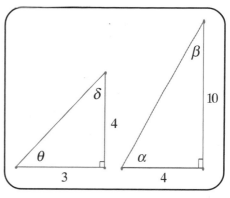

Figure 2.7.

to use the Pythagorean Theorem to calculate the length of the hypotenuse of each triangle in order to evaluate these ratios. For the first triangle, $3^2 + 4^2 = c^2$, so the hypotenuse is 5. For the second triangle, $4^2 + 10^2 = c^2$, so the hypotenuse is $2\sqrt{29}$. Using the equations for the sine and cosine, we see that:

$$\sin\theta = \frac{4}{5} \quad \sin\delta = \frac{3}{5} \quad \sin\alpha = \frac{10}{2\sqrt{29}} = \frac{5\sqrt{29}}{29} \quad \sin\beta = \frac{4}{2\sqrt{29}} = \frac{2\sqrt{29}}{29}$$

$$\cos\theta = \frac{3}{5} \quad \cos\delta = \frac{4}{5} \quad \cos\alpha = \frac{4}{2\sqrt{29}} = \frac{2\sqrt{29}}{29} \quad \cos\beta = \frac{10}{2\sqrt{29}} = \frac{5\sqrt{29}}{29}$$

There are some interesting observations to make about the calculations in Example 1. First of all, notice that $\sin\theta = \cos\delta$ and $\sin\delta = \cos\theta$. Also, $\sin\beta = \cos\alpha$ and $\sin\alpha = \cos\beta$. Remember that the interior angles of a triangle add up to 180°, and because one of the angles in each of the right triangles measures 90°, the angles θ, δ, α, and β are related by the equations $\theta = 90 - \delta$ and $\beta = 90 - \alpha$. If we consider the sine and cosine of an angle as functions of the angle θ, then the sine and cosine functions have the relationship $\sin\theta = \cos(90° - \theta)$. This relationship can also be stated when the angle measurement is in radians: $\sin\theta = \cos\left(\frac{\pi}{2} - \theta\right)$.

There are a few special triangles and angles that are worth knowing very well. An isosceles right triangle, as shown in Figure 2.8, is one in

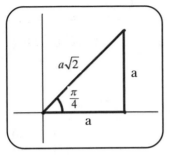

Figure 2.8.

which the two acute angles of the triangle have the same measure: 45°. A 45° angle corresponds to an angle with measure $\frac{\pi}{4}$ radians. In an isosceles right triangle, both legs have the same length, a, and the hypotenuse has length $a\sqrt{2}$.

We see that $\sin\frac{\pi}{4} = \frac{a}{a\sqrt{2}} = \frac{1}{\sqrt{2}} = \frac{\sqrt{2}}{2}$ and

$$\cos\frac{\pi}{4} = \frac{a}{a\sqrt{2}} = \frac{1}{\sqrt{2}} = \frac{\sqrt{2}}{2}.$$

Another triangle worth being familiar with is a 30° − 60° − 90° triangle, as shown in Figure 2.9. In this triangle, if the shorter of the two legs has length a, then the hypotenuse has length $2a$ and the length of the longer leg is $a\sqrt{3}$. The longer leg is opposite the larger acute angle, and the hypotenuse is opposite the right angle. An angle whose measure is 30° has an equivalent measure of $\frac{\pi}{6}$ radians, and an angle whose measure is 60° has an equivalent measure of $\frac{\pi}{3}$ radians. We can determine the sine and cosine of these angles by using the lengths of the sides of the right triangle:

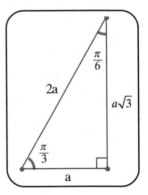

Figure 2.9.

$$\sin\frac{\pi}{6} = \frac{a}{2a} = \frac{1}{2}$$

$$\cos\frac{\pi}{6} = \frac{a\sqrt{3}}{2a} = \frac{\sqrt{3}}{2}$$

$$\sin\frac{\pi}{3} = \frac{a\sqrt{3}}{2a} = \frac{\sqrt{3}}{2}$$

$$\cos\frac{\pi}{3} = \frac{a}{2a} = \frac{1}{2}$$

Even though there are some angles that do not fit into a right triangle directly (for example, angles with measure 0 or right angles themselves) interpreting the definitions of the sine and cosine of an angle loosely will enable us to determine meaningful and consistent values for the sine and cosine of these particular angles. We will start with an angle of measure $\frac{\pi}{2}$ radians (or 90°). A right triangle has an angle of measure $\frac{\pi}{2}$ radians, and the side opposite the right angle is, in fact, the hypotenuse. Using this information, we can evaluate $\sin\frac{\pi}{2}$:

$$\sin\frac{\pi}{2} = \frac{\text{opposite}}{\text{hypotenuse}} = \frac{\text{hypotenuse}}{\text{hypotenuse}} = 1$$

Combining this with our relationship $\sin\theta = \cos\left(\frac{\pi}{2} - \theta\right)$ enables us to evaluate cos 0 by substituting $\theta = \frac{\pi}{2}$ into this equation:

$$\sin\theta = \cos\left(\frac{\pi}{2} - \theta\right)$$
$$\sin\frac{\pi}{2} = \cos\left(\frac{\pi}{2} - \frac{\pi}{2}\right)$$
$$\sin\frac{\pi}{2} = \cos 0$$
$$\cos 0 = 1$$

To evaluate sin 0, imagine a triangle with an angle of measure 0 (in radians or degrees). The triangle would have no height (and technically it would not be a triangle at all, but the idea helps us understand what sin 0 *should* be). The length of the side opposite this angle of measure 0 is 0, which means that:

$$\sin 0 = \frac{\text{opposite}}{\text{hypotenuse}} = \frac{0}{\text{hypotenuse}} = 0$$

This value for sin 0 will enable us to find $\cos\frac{\pi}{2}$ by substituting $\theta = 0$ into the equation $\sin\theta = \cos\left(\frac{\pi}{2} - \theta\right)$:

$$\sin\theta = \cos\left(\frac{\pi}{2} - \theta\right)$$
$$\sin 0 = \cos\left(\frac{\pi}{2} - 0\right)$$
$$\cos\frac{\pi}{2} = 0$$

Keep in mind that if you understand the sine ratio, you gain insight into the cosine ratio through the relationship $\sin\theta = \cos\left(\frac{\pi}{2} - \theta\right)$. The values of sin θ and cos θ for the angles that we have discussed, specifically 0, $\frac{\pi}{6}$, $\frac{\pi}{4}$, $\frac{\pi}{3}$, and $\frac{\pi}{2}$, are important enough for me to summarize in the following table. I recommend that you become very familiar with the sine and cosine of these very special angles.

θ		$\sin\theta$	$\cos\theta$
0	0°	0	1
$\frac{\pi}{6}$	30°	$\frac{1}{2}$	$\frac{\sqrt{3}}{2}$
$\frac{\pi}{4}$	45°	$\frac{\sqrt{2}}{2}$	$\frac{\sqrt{2}}{2}$
$\frac{\pi}{3}$	60°	$\frac{\sqrt{3}}{2}$	$\frac{1}{2}$
$\frac{\pi}{2}$	90°	1	0

We will add to this table as we learn about the other trigonometric ratios.

There are six trigonometric ratios. The first two that we have defined are the sine and cosine ratios. These two ratios are the building blocks to the other four trigonometric ratios that we will define. The **tangent** of an angle is the ratio of the sine of the angle to the cosine of the angle:

$$\tan\theta = \frac{\sin\theta}{\cos\theta}$$

We can use the definitions of the sine and cosine ratios to understand the tangent ratio.

Remember that $\sin\theta = \dfrac{\text{opposite}}{\text{hypotenuse}}$, and

$\cos\theta = \dfrac{\text{adjacent}}{\text{hypotenuse}}$. Taking the ratio of these

two functions gives:

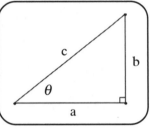

Figure 2.10.

$$\tan\theta = \frac{\sin\theta}{\cos\theta} = \frac{\dfrac{\text{opposite}}{\text{hypotenuse}}}{\dfrac{\text{adjacent}}{\text{hypotenuse}}} = \frac{\text{opposite}}{\text{hypotenuse}} \times \frac{\text{hypotenuse}}{\text{adjacent}} = \frac{\text{opposite}}{\text{adjacent}}$$

In terms of the right triangle shown in Figure 2.10, $\tan\theta = \dfrac{b}{a}$.

Example 2

Find the sine, cosine, and tangent of the angles α and β, as shown in Figure 2.11.

Solution: Use the Pythagorean Theorem to find the length of the third side of each triangle. Then use the definitions of the sine, cosine, and tangent ratios.

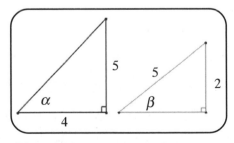

Figure 2.11.

a. The length of the hypotenuse is $\sqrt{4^2 + 5^2} = \sqrt{16 + 25} = \sqrt{41}$.
   From this, we have:

$$\sin\alpha = \frac{4}{\sqrt{41}} \qquad \cos\alpha = \frac{5}{\sqrt{41}} \qquad \tan\alpha = \frac{4}{5}$$

b. The length of the leg of the other triangle is

$\sqrt{5^2 - 2^2} = \sqrt{25 - 4} = \sqrt{21}$. From this, we have:

$$\sin\beta = \frac{2}{5} \qquad \cos\beta = \frac{\sqrt{21}}{5} \qquad \tan\beta = \frac{2}{\sqrt{21}}$$

The tangent of an angle will be defined as long as the cosine of the angle is not equal to 0. We already know of one angle whose cosine is 0: $\cos\frac{\pi}{2} = 0$. When we examine the cyclic nature of the trigonometric functions, we will see that there are many more angles whose cosine is 0.

Example 3

Find the tangent of the special angles 0, $\frac{\pi}{6}$, $\frac{\pi}{4}$, $\frac{\pi}{3}$, and $\frac{\pi}{2}$.

Solution: Using the equation $\tan\theta = \dfrac{\sin\theta}{\cos\theta}$, we see that:

$$\tan 0 = \frac{\sin 0}{\cos 0} = \frac{0}{1} = 0$$

$$\tan\frac{\pi}{6} = \frac{\sin\frac{\pi}{6}}{\cos\frac{\pi}{6}} = \frac{\frac{1}{2}}{\frac{\sqrt{3}}{2}} = \frac{1}{\sqrt{3}} = \frac{\sqrt{3}}{3}$$

$$\tan\frac{\pi}{4} = \frac{\sin\frac{\pi}{4}}{\cos\frac{\pi}{4}} = \frac{\frac{\sqrt{2}}{2}}{\frac{\sqrt{2}}{2}} = 1$$

$$\tan\frac{\pi}{3} = \frac{\sin\frac{\pi}{3}}{\cos\frac{\pi}{3}} = \frac{\frac{\sqrt{3}}{2}}{\frac{1}{2}} = \sqrt{3}$$

$$\tan\frac{\pi}{2} = \frac{\sin\frac{\pi}{2}}{\cos\frac{\pi}{2}} = \frac{1}{0} = \text{undefined}$$

θ		$\sin\theta$	$\cos\theta$	$\tan\theta$
0	0°	0	1	0
$\frac{\pi}{6}$	30°	$\frac{1}{2}$	$\frac{\sqrt{3}}{2}$	$\frac{\sqrt{3}}{3}$
$\frac{\pi}{4}$	45°	$\frac{\sqrt{2}}{2}$	$\frac{\sqrt{2}}{2}$	1
$\frac{\pi}{3}$	60°	$\frac{\sqrt{3}}{2}$	$\frac{1}{2}$	$\sqrt{3}$
$\frac{\pi}{2}$	90°	1	0	undefined

The special angles that we studied in the last lesson continue to be important. I will include the values of the tangent of these angles in the table we started earlier.

The last three trigonometric ratios are the reciprocals of the first three trigonometric ratios. We define the secant of an angle in terms of the cosine of the angle:

$$\sec\theta = \frac{1}{\cos\theta}$$

The cosecant of an angle is the reciprocal of the sine of the angle:

$$\csc\theta = \frac{1}{\sin\theta}$$

The cotangent of an angle is the reciprocal of the tangent of the angle:

$$\cot\theta = \frac{1}{\tan\theta}$$

Because the tangent of an angle is itself related to the sine and cosine of the angle, we can also write the cotangent of an angle in terms of the sine and cosine of the angle:

$$\cot\theta = \frac{1}{\tan\theta} = \frac{1}{\frac{\sin\theta}{\cos\theta}} = \frac{\cos\theta}{\sin\theta}$$

There are many ways to write the formulas for the reciprocal trigonometric ratios. Because $\csc\theta = \frac{1}{\sin\theta}$, we can write the cotangent of an angle as:

$$\cot\theta = \frac{\cos\theta}{\sin\theta} = \cos\theta\csc\theta$$

The trigonometric ratios are all related to each other, and we will examine some of these relationships in more detail in Chapter 5.

The secant of an angle will be defined as long as the cosine of the angle is not equal to 0. The cosecant of an angle will be defined as long as the sine of the angle is not equal to 0. Finally, the cotangent of an angle will be defined as long as the sine of the angle is not equal to 0.

We can evaluate the secant, cosecant, and cotangent of the angles 0, $\frac{\pi}{6}$, $\frac{\pi}{4}$, $\frac{\pi}{3}$, and $\frac{\pi}{2}$, and expand our table. Again, I would recommend that you become familiar with the trigonometric ratios of these special angles.

θ		$\sin \theta$	$\cos \theta$	$\tan \theta$	$\csc \theta$	$\sec \theta$	$\cot \theta$
0	0°	0	1	0	undefined	1	undefined
$\frac{\pi}{6}$	30°	$\frac{1}{2}$	$\frac{\sqrt{3}}{2}$	$\frac{\sqrt{3}}{3}$	2	$\frac{2\sqrt{3}}{3}$	$\sqrt{3}$
$\frac{\pi}{4}$	45°	$\frac{\sqrt{2}}{2}$	$\frac{\sqrt{2}}{2}$	1	$\sqrt{2}$	$\sqrt{2}$	1
$\frac{\pi}{3}$	60°	$\frac{\sqrt{3}}{2}$	$\frac{1}{2}$	$\sqrt{3}$	$\frac{2\sqrt{3}}{3}$	2	$\frac{\sqrt{3}}{3}$
$\frac{\pi}{2}$	90°	1	0	undefined	1	undefined	0

The names of the trigonometric ratios are significant. The trigonometric ratios can be paired together: sine and cosine, tangent and cotangent, secant and cosecant. These pairs are sometimes referred to as **cofunctions**. Earlier we observed that $\sin\theta = \cos\left(\frac{\pi}{2} - \theta\right)$. A generalization of this observation is called the *Complementary Angle Theorem*.

The Complementary Angle Theorem:
Cofunctions of complementary angles are equal.

This means that in addition to the relationship $\sin\theta = \cos\left(\frac{\pi}{2} - \theta\right)$, we have $\sec\theta = \csc\left(\frac{\pi}{2} - \theta\right)$ and $\tan\theta = \cot\left(\frac{\pi}{2} - \theta\right)$. The following equations illustrate the Complementary Angle Theorem:

$$\sin 40° = \cos\left(90° - 40°\right) = \cos 50°$$

$$\tan\frac{\pi}{8} = \cot\left(\frac{\pi}{2} - \frac{\pi}{8}\right) = \cot\frac{3\pi}{8}$$

$$\sec\frac{\pi}{6} = \csc\left(\frac{\pi}{2} - \frac{\pi}{6}\right) = \csc\frac{\pi}{3}$$

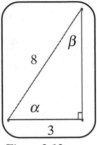
Figure 2.12.

Lesson 2-3 Review

1. Find the sine, cosine, tangent, secant, cosecant, and cotangent of the angles α and β as shown in Figure 2.12:

2. If $\sec \theta = 4$, find $\sin \theta$, $\cos \theta$, $\tan \theta$, $\csc \theta$, and $\cot \theta$.

3. If $\tan \theta = 5$, find the exact value of

$$\tan \theta + \tan\left(\frac{\pi}{2} - \theta\right).$$

Lesson 2-4: Trigonometric Ratios of General Angles

The trigonometric ratios are actually *functions* whose input is an angle and whose output is a ratio of two lengths. The units for the input can be either radians or degrees, and the output has no units, or is **dimensionless**. The development of the sine and cosine ratios using right triangles has one drawback: we can only evaluate the trigonometric ratios of acute angles. To extend the definitions of the trigonometric functions to angles that are not acute, we will orient the angle in standard position. Remember that an angle is in standard position if its vertex is located at the origin of the Cartesian coordinate system and its initial side coincides with the positive x-axis. If θ is an angle in standard position, let (a, b) denote *any* point on the terminal side of θ, other than the origin. We will let r represent the distance between (a, b) and the origin:

$$r = \sqrt{a^2 + b^2}$$

Then the six trigonometric functions of θ are defined as follows:

$$\sin \theta = \frac{b}{r} \qquad \cos \theta = \frac{a}{r} \qquad \tan \theta = \frac{b}{a} \text{ if } a \neq 0$$

$$\csc \theta = \frac{r}{b} \text{ if } b \neq 0 \qquad \sec \theta = \frac{r}{a} \text{ if } a \neq 0 \qquad \cot \theta = \frac{a}{b} \text{ if } b \neq 0$$

If $a = 0$, then $\tan \theta$ and $\sec \theta$ are not defined. If $b = 0$, then $\cot \theta$ and $\csc \theta$ are not defined.

Surprisingly enough, the values of the six trigonometric functions of an angle θ are independent of the point on the terminal side that is used.

The trigonometric functions only depend on the angle θ. This is because of the properties of similar triangles. Suppose that (a, b) and (c, d) are points on the terminal side of an angle θ in standard position, as shown in Figure 2.13. Then the triangles $\triangle OPQ$ and $\triangle ORS$ are similar triangles, and corresponding sides of similar triangles are proportional. For example, $\frac{a}{b} = \frac{c}{d}$, and it does not matter whether we write $\tan\theta = \frac{a}{b}$ or $\tan\theta = \frac{c}{d}$. The value of tan θ is the same regardless of which ratio we use.

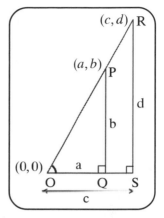

Figure 2.13.

Now that we are looking at the trigonometric ratios as *functions*, we will need to determine their domains. From the definition of the sine and cosine functions,

$$\sin\theta = \frac{b}{r} \text{ and } \cos\theta = \frac{a}{r}$$

we can see that, as long as the point on the terminal side of θ is not the origin, it will be possible to evaluate the sine and cosine of any angle. The domain sin θ and cos θ is the set of all real numbers. If the point on the terminal side of θ is in Quadrant I, then $a > 0$ and $b > 0$. From this, we see that if θ lies in Quadrant I, sin $\theta > 0$ and cos $\theta > 0$. If the point on the terminal side of θ is in Quadrant II, then $a < 0$ and $b > 0$. From this, we see that if θ lies in Quadrant II, sin $\theta > 0$ and cos $\theta < 0$. If the point on the terminal side of θ is in Quadrant III, then $a < 0$ and $b < 0$. From this, we see that if θ lies in Quadrant III, sin $\theta < 0$ and cos $\theta < 0$. If the point on the terminal side of θ is in Quadrant IV, then $a > 0$ and $b < 0$. From this, we see that if θ lies in Quadrant IV, sin $\theta < 0$ and cos $\theta > 0$.

We can summarize these observations. If the point (a, b) lies in Quadrants I or II, the sine of the angle formed will be positive. If (a, b) lies in Quadrants III or IV, then the sine of the angle formed will be negative. Similarly, the cosine of the angle will be positive if the x-coordinate of (a, b) is positive. In other words, if (a, b) lies in Quadrants I or IV, the cosine of the angle formed will be positive, and if (a, b) lies in quadrants II or III, the cosine of the angle formed will be negative. The signs of the sine and the cosine functions for the four quadrants are listed in the table following on page 42.

The pneumonic **All Students Take Calculus** may help you remember which trigonometric functions are positive in which quadrants. In Quadrant I, all of the trigonometric functions are positive. In Quadrant II, the sine ratio is positive, in Quadrant III, the tangent ratio is positive. In Quadrant IV, the cosine ratio is positive.

Quadrant	Sign of the Sine	Sign of the Cosine
I	+	+
II	+	−
III	−	−
IV	−	+

Because the other trigonometric functions are defined in terms of the sine and cosine functions, once you know the signs of the sine and cosine functions, you know the signs of the other four trigonometric functions.

Example 1

Find the exact values of the six trigonometric functions of θ corresponding to the following points on the terminal side of θ:

a. (3, 4)

c. (−5, 12)

b. (4, −3)

d. (1, −1)

Solution: Find the values of a, b, and r, and use the formulas for the trigonometric functions:

$$\sin\theta = \frac{b}{r} \qquad \cos\theta = \frac{a}{r} \qquad \tan\theta = \frac{b}{a} \text{ if } a \neq 0$$

$$\csc\theta = \frac{r}{b} \text{ if } b \neq 0 \qquad \sec\theta = \frac{r}{a} \text{ if } a \neq 0 \qquad \cot\theta = \frac{a}{b} \text{ if } b \neq 0$$

a. $a = 3, b = 4, r = \sqrt{3^2 + 4^2} = \sqrt{9+16} = \sqrt{25} = 5$

$$\sin\theta = \frac{4}{5} \qquad \cos\theta = \frac{3}{5} \qquad \tan\theta = \frac{4}{3}$$

$$\csc\theta = \frac{5}{4} \qquad \sec\theta = \frac{5}{3} \qquad \cot\theta = \frac{3}{4}$$

b. $a = 4, b = -3, \quad r = \sqrt{4^2 + (-3)^2} = \sqrt{16 + 9} = \sqrt{25} = 5$

$$\sin\theta = -\frac{3}{5} \qquad \cos\theta = \frac{4}{5} \qquad \tan\theta = -\frac{3}{4}$$

$$\csc\theta = -\frac{5}{3} \qquad \sec\theta = \frac{5}{4} \qquad \cot\theta = -\frac{4}{3}$$

c. $a = -5, b = 12, \quad r = \sqrt{(-5)^2 + 12^2} = \sqrt{25 + 144} = \sqrt{169} = 13$

$$\sin\theta = \frac{12}{13} \qquad \cos\theta = -\frac{5}{13} \qquad \tan\theta = -\frac{12}{5}$$

$$\csc\theta = \frac{13}{12} \qquad \sec\theta = -\frac{13}{5} \qquad \cot\theta = -\frac{5}{12}$$

d. $a = 1, b = -1, \quad r = \sqrt{1^2 + (-1)^2} = \sqrt{1 + 1} = \sqrt{2}$

$$\sin\theta = -\frac{1}{\sqrt{2}} = -\frac{\sqrt{2}}{2} \qquad \cos\theta = \frac{1}{\sqrt{2}} = \frac{\sqrt{2}}{2} \qquad \tan\theta = -\frac{1}{1} = -1$$

$$\csc\theta = -\frac{\sqrt{2}}{1} = -\sqrt{2} \qquad \sec\theta = \frac{\sqrt{2}}{1} = \sqrt{2} \qquad \cot\theta = -\frac{1}{1} = -1$$

Follow this same procedure when evaluating the trigonometric functions at the quadrantal angles. At each of the quadrantal angles, either $a = 0$ and $b = \pm r$ (meaning that $\cos\theta = 0$ and $\sin\theta = \pm 1$) or $b = 0$ and $a = \pm r$ (meaning that $\sin\theta = 0$ and $\cos\theta = \pm 1$). Keep in mind that two of the six trigonometric functions will not be defined if the terminal side of an angle coincides with a coordinate axis, resulting in a quadrantal angle. For the quadrantal angles that are integer multiples of π ($\dots, -2\pi, -\pi, 0, \pi, 2\pi, \dots$), $b = 0$ and $\csc\theta$ and $\cot\theta$ will not be defined. For the quadrantal angles that are odd half-integer multiples of π ($\dots, -\frac{3\pi}{2}, -\frac{\pi}{2}, \frac{\pi}{2}, \frac{3\pi}{2}, \dots$), $a = 0$ so $\sec\theta$ and $\tan\theta$ will not be defined. The values of the trigonometric functions at the quadrantal angles are summarized in the table on page 44.

θ (radians)	θ (degrees)	$\sin \theta$	$\cos \theta$	$\tan \theta$	$\csc \theta$	$\sec \theta$	$\cot \theta$
0	0°	0	1	0	undefined	1	undefined
$\frac{\pi}{2}$	90°	1	0	undefined	1	undefined	0
π	180°	0	−1	0	undefined	−1	undefined
$\frac{3\pi}{2}$	270°	−1	0	undefined	−1	undefined	0

In order to work with angles that correspond to measures that are greater than 360° (or 2π radians), or are negative, we need to introduce the concept of coterminal angles. Two angles in standard position are **coterminal** if they have the same terminal side. For example, angles with measure 60° and −300° are coterminal, as are the pair of angles measuring 50° and 410°. Figure 2.14 shows some examples of coterminal angles.

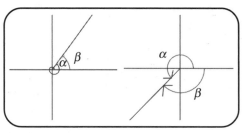

In trigonometry, a picture is worth a thousand words. When evaluating a trigonometric function at a particular angle, it is often helpful to draw the angle in standard position.

Figure 2.14.

Example 2

Find the exact values of the following:

a. $\tan (-315°)$
b. $\csc \frac{5\pi}{2}$
c. $\cos 390°$

Solution: Sketch the angles first, and then use the table of trigonometric functions of special angles that we developed earlier in this chapter.

a. $\tan (-315°)$: The angle is shown in Figure 2.15. An angle of measure −315° is coterminal with an angle of measure 45°, so $\tan (-315°) = \tan 45° = 1$.

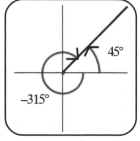

Figure 2.15.

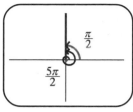

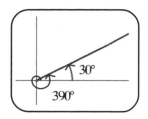

Figure 2.16.　　　　　　　　*Figure 2.17.*

b. $\csc\frac{5\pi}{2}$: The angle is shown in Figure 2.16. An angle of measure $\frac{5\pi}{2}$

is coterminal with an angle of measure $\frac{\pi}{2}$, so $\csc\frac{5\pi}{2} = \csc\frac{\pi}{2} = 1$.

c. cos 390°: The angle is shown in Figure 2.17. An angle of measure 390° is coterminal with an angle of measure 30°, so

$$\cos 390° = \cos 30° = \frac{\sqrt{3}}{2}.$$

Coterminal angles can be used to evaluate trigonometric functions for angles that are coterminal with acute angles or quadrantal angles. We still need a method for evaluating the trigonometric functions for angles that are not in Quadrant I or are quadrantal angles. We already know the *signs* of the trigonometric functions for angles that are in all four Quadrants. Now we need to determine the *magnitude*. In order to do this, we will introduce the concept of a *reference angle*. The *reference angle* of an angle θ is an *acute* angle α with the property that:

$\lvert \sin\theta \rvert = \sin\alpha$	$\lvert \cos\theta \rvert = \cos\alpha$	$\lvert \tan\theta \rvert = \tan\alpha$
$\lvert \csc\theta \rvert = \csc\alpha$	$\lvert \sec\theta \rvert = \sec\alpha$	$\lvert \cot\theta \rvert = \cot\alpha$

Let θ represent a non-acute angle that lies in the coordinate plane. The acute angle formed by the terminal side of θ and either the positive *x*-axis or the negative *x*-axis is called the **reference angle** for θ. A reference angle is always an acute angle. If θ lies in Quadrant II or Quadrant III, the acute angle will be formed by the terminal side of θ and the negative *x*-axis. If θ lies in Quadrant IV, the acute angle will be formed by the terminal side of θ and the positive *x*-axis. If θ lies in Quadrant I, the reference angle is the acute angle that is coterminal with θ. The easiest way to determine the reference angle of an angle θ is to draw a picture.

Example 3

Find the reference angle for each of the following angles:

a. 300°   c. −100°

b. $\frac{7\pi}{4}$ d. $-\frac{4\pi}{3}$

Solution: Draw each of the angles and use your drawing to determine the reference angle.

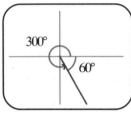

Figure 2.18.

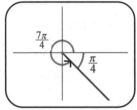

Figure 2.19.

a. 300°: The angle is shown in Figure 2.18. The reference angle of an angle with measure 300° has measure 60°.

b. $\frac{7\pi}{4}$: The angle is shown in Figure 2.19. The reference angle of an angle with measure $\frac{7\pi}{4}$ has measure $\frac{\pi}{4}$.

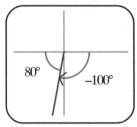

Figure 2.20.

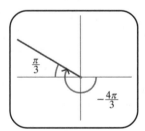

Figure 2.21.

c. −100°: The angle is shown in Figure 2.20. The reference angle of an angle with measure −100° has measure 80°.

d. $-\frac{4\pi}{3}$: The angle is shown in Figure 2.21. The reference angle of an angle with measure $-\frac{4\pi}{3}$ has measure $\frac{\pi}{3}$.

Reference angles are used to evaluate trigonometric functions of general angles. If θ is an angle in standard position and has a reference angle α, then the *magnitude* of the trigonometric functions at θ and α are the same. The signs of the trigonometric functions at θ are determined by the quadrant that θ lies in.

$\sin \theta = \pm\sin \alpha$	$\cos \theta = \pm\cos \alpha$	$\tan \theta = \pm\tan \alpha$
$\csc \theta = \pm\csc \alpha$	$\sec \theta = \pm\sec \alpha$	$\cot \theta = \pm\cot \alpha$

Example 4

Find the exact value of the following:

a. $\sin 300°$

c. $\cos 780°$

b. $\tan \dfrac{7\pi}{4}$

d. $\sec\left(-\dfrac{2\pi}{3}\right)$

Solution: Determine the reference angle and the quadrant that the angle lies in.

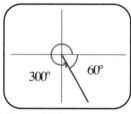

Figure 2.22.

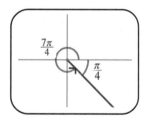

Figure 2.23.

a. $\sin 300°$: The angle is shown in Figure 2.22. It lies in Quadrant IV and its reference angle has measure $60°$. The sine of an angle in Quadrant IV is negative, so $\sin 300° = -\sin 60° = -\dfrac{\sqrt{3}}{2}$.

b. $\tan \dfrac{7\pi}{4}$: The angle is shown in Figure 2.23. It lies in Quadrant IV and its reference angle has measure $\dfrac{\pi}{4}$. The tangent of an angle in Quadrant IV is negative, so $\tan \dfrac{7\pi}{4} = -\tan \dfrac{\pi}{4} = -1$.

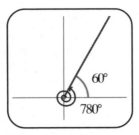
Figure 2.24.

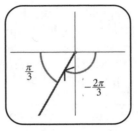
Figure 2.25.

c. cos 780°: The angle is shown in Figure 2.24. It lies in Quadrant I and its reference angle has measure 60°. The cosine of an angle in Quadrant I is positive, so $\cos 780° = \cos 60° = \frac{1}{2}$.

d. $\sec\left(-\frac{2\pi}{3}\right)$: The angle is shown in Figure 2.25. It lies in Quadrant III and its reference angle has measure $\frac{\pi}{3}$. The secant of an angle in Quadrant III is negative, so $\sec\left(-\frac{2\pi}{3}\right) = -\sec\left(\frac{\pi}{3}\right) = -2$.

Example 5

If tan θ = –2 and cos θ < 0, find the exact value of the remaining trigonometric functions evaluated at θ.

Solution: The only quadrant where both the tangent and cosine of an angle are negative is Quadrant II. Therefore, θ must lie in Quadrant II. If α is the reference angle for θ, then tan α = 2. Draw a right triangle and label the lengths of two of the sides using the fact that tan α = 2, as shown in Figure 2.26. You are making use of the *right triangle* definition of the tangent function:

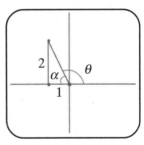
Figure 2.26.

$$\tan\alpha = \frac{\text{opposite}}{\text{adjacent}}$$

Use the Pythagorean Theorem to determine the length of the hypotenuse. From this, we have:

$$\sin\theta = \sin\alpha = \frac{2}{\sqrt{5}} \qquad \cos\theta = -\cos\alpha = -\frac{1}{\sqrt{5}} \qquad \tan\theta = -2$$

$$\csc\theta = \csc\alpha = \frac{\sqrt{5}}{2} \qquad \sec\theta = -\sec\alpha = -\sqrt{5} \qquad \cot\theta = -\frac{1}{2}$$

In order to determine the value of a trigonometric function of a general (non-quadrantal) angle θ, follow the following procedure:

1. Determine the quadrant that θ lies in. This will determine the sign of the trigonometric function.

2. Determine the reference angle, α, of θ. This will enable you to determine the magnitude of the trigonometric function.

3. Use the right triangle definitions of the trigonometric functions to determine the lengths of two of the three sides of the triangle, depending on which trigonometric function value you are given:

$$\sin\alpha = \frac{\text{opposite}}{\text{hypotenuse}} \qquad \cos\alpha = \frac{\text{adjacent}}{\text{hypotenuse}} \qquad \tan\alpha = \frac{\text{opposite}}{\text{adjacent}}$$

$$\csc\alpha = \frac{\text{hypotenuse}}{\text{opposite}} \qquad \sec\alpha = \frac{\text{hypotenuse}}{\text{adjacent}} \qquad \cot\alpha = \frac{\text{adjacent}}{\text{opposite}}$$

4. Use the Pythagorean Theorem to determine the length of the third side.

5. Once you have determined the lengths of all three sides of the right triangle, you can find the value of any of the trigonometric functions at α, and hence the magnitude of the trigonometric functions at θ.

6. Based on the quadrant of θ, determine the signs of the trigonometric functions, and combine this information with the magnitude of the trigonometric functions to obtain the values of the trigonometric functions at θ.

Lesson 2-4 Review

1. Find the exact values of the six trigonometric functions of θ if the point $(-2, 5)$ lies on the terminal side of θ.

2. Find the exact value of $\tan\left(-\frac{5\pi}{4}\right)$.

3. If csc $\theta = -4$ and tan $\theta > 0$, find the exact value of the remaining trigonometric functions evaluated at θ.

Lesson 2-5: The Pythagorean Identities

We can use the formulas for the trigonometric functions to establish some important identities. A **trigonometric identity** is an equation that is satisfied for all values of the variable. The equation $(x + 1)^2 = x^2 + 2x + 1$ is an example of an **identity**, or an equation that is true regardless of the value of x used. Equations that are satisfied only for *particular* values of the variable are called **conditional** equations. The equation $x + 1 = 0$ is an example of a conditional equation, because the equation is only satisfied when $x = -1$.

The first trigonometric identity we will establish originates from the definition of the sine and cosine functions. We can square both sides of the equation $r = \sqrt{a^2 + b^2}$ to get the equation $a^2 + b^2 = r^2$. If we divide both sides of this equation by r^2, we get:

$$\frac{a^2}{r^2} + \frac{b^2}{r^2} = 1$$

or:

$$\left(\frac{a}{r}\right)^2 + \left(\frac{b}{r}\right)^2 = 1$$

We can use the formulas for the sine and cosine functions to rewrite this equation:

$(\cos \theta)^2 + (\sin \theta)^2 = 1$

It is awkward to use parentheses when we write $(\cos \theta)^2$ and $(\sin \theta)^2$, so we will shorten our notation. Write $(\cos \theta)^2$ as $\cos^2 \theta$ and $(\sin \theta)^2$ as $\sin^2 \theta$. Our first trigonometric identity is:

$\cos^2 \theta + \sin^2 \theta = 1$

Because this is an identity, this equation holds for every angle θ.

From this first identity, we can easily derive a second identity. Divide both sides of this equation by $\cos^2 \theta$ and simplify:

$\cos^2 \theta + \sin^2 \theta = 1$

$$\frac{\cos^2 \theta}{\cos^2 \theta} + \frac{\sin^2 \theta}{\cos^2 \theta} = \frac{1}{\cos^2 \theta}$$

$1 + \tan^2 \theta = \sec^2 \theta$

This second identity can be surprisingly helpful and is easy to derive from the first identity. There are many more trigonometric identities that establish various relationships between the trigonometric functions, and we will explore more of these trigonometric identities in detail in Chapter 5.

Answer Key
Lesson 2-1 Review

1. Find the angular speed and then use the equation $v = r\omega$ to find the linear speed:

$$\omega = \frac{120 \; \text{revolutions}}{\text{minute}} \cdot \frac{2\pi \; \text{radians}}{\text{revolution}} = \frac{240\pi \; \text{radians}}{\text{minute}}$$

$$v = r\omega = (2 \; \text{feet}) \cdot \frac{240\pi \; \text{radians}}{\text{minute}} = \frac{480\pi \; \text{feet}}{\text{minute}}$$

The linear speed will be $\dfrac{480\pi \; \text{feet}}{\text{minute}}$, or approximately 1508 $^{\text{ft}}\!/_{\text{min}}$.

2. Use the formulas for the arc length and area of a sector. Remember to use the radius of the circle, not the diameter!

$s = r\theta = (5)\left(\frac{100\pi}{180}\right) = \frac{25\pi}{9} \; \text{cm}$

$A = \frac{1}{2}r^2\theta = \frac{1}{2}(5)^2 \left(\frac{100\pi}{180}\right) = \frac{1}{2}(25)\left(\frac{100\pi}{180}\right) = \frac{125\pi}{18} \; \text{cm}^2$

3. Use the formulas for the arc length and the area of a sector together to first find the radius of the circle. Then use the formula for the arc length of a sector of a circle to solve for the central angle:

$2\pi = r\theta$ and $10\pi = \frac{1}{2}r^2\theta = \frac{1}{2}(r\theta)r$

$10\pi = \frac{1}{2}(2\pi)r$

$10\pi = \pi r$

$r = 10$

$2\pi = 10\theta$

$\theta = \frac{\pi}{5}$

Lesson 2-3 Review

1. The length of the other leg of the triangle is $\sqrt{55}$, so $\sin\alpha = \cos\beta = \frac{3}{8}$,

    $\cos\alpha = \sin\beta = \frac{\sqrt{55}}{8}$, $\tan\alpha = \cot\beta = \frac{\sqrt{55}}{3}$, $\sec\alpha = \csc\beta = \frac{8}{3}$,

    $\csc\alpha = \sec\beta = \frac{8}{\sqrt{55}}$, and $\cot\alpha = \tan\beta = \frac{3}{\sqrt{55}}$.

2. If $\sec\theta = 4$, construct a right triangle whose hypotenuse has length 4 and the adjacent leg has length 1. By the Pythagorean Theorem, the length of the

    opposite leg is $\sqrt{15}$. From this, we see that $\cos\theta = \frac{1}{4}$, $\sin\theta = \frac{\sqrt{15}}{4}$,

    $\tan\theta = \sqrt{15}$, $\csc\theta = \frac{4}{\sqrt{15}}$, and $\cot\theta = \frac{1}{\sqrt{15}}$.

3. If $\tan\theta = 5$, $\tan\theta + \tan\left(\frac{\pi}{2} - \theta\right) = \tan\theta + \cot\theta = 5 + \frac{1}{5} = \frac{26}{25}$.

Lesson 2-4 Review

1. θ lies in Quadrant II. $\sin\theta = \frac{5}{\sqrt{29}}$, $\cos\theta = -\frac{2}{\sqrt{29}}$, $\tan\theta = -\frac{5}{2}$, $\csc\theta = \frac{\sqrt{29}}{5}$,

    $\sec\theta = -\frac{\sqrt{29}}{2}$, and $\cot\theta = -\frac{2}{5}$.

2. $\tan\left(-\frac{5\pi}{4}\right) = -1$

3. If $\csc\theta = -4$ and $\tan\theta > 0$, then θ lies in Quadrant III, so only the tangent and cotangent functions will be positive. Construct a right triangle whose hypotenuse has length 4 and the opposite leg has length 1. By the Pythagorean Theorem, the length of the adjacent leg is $\sqrt{15}$. From this, we see that

    $\sin\theta = -\frac{1}{4}$, $\cos\theta = -\frac{\sqrt{15}}{4}$, $\tan\theta = \frac{1}{\sqrt{15}}$, $\sec\theta = -\frac{4}{\sqrt{15}}$, and $\cot\theta = \sqrt{15}$.

Trigonometric Functions

In Chapter 2, we developed the trigonometric functions using the right triangle approach. This approach makes use of concepts that are familiar to us: right triangles and the Pythagorean Theorem. With this approach, we can evaluate the trigonometric functions for particular angles, but the periodic nature of these functions is not as apparent. The unit circle approach, on the other hand, emphasizes the periodic nature of these functions. Both approaches have their advantages, as we will see.

Lesson 3-1: The Unit Circle Approach

We can define the sine and cosine of an angle of *any* measure using a unit circle. Remember that the unit circle is a circle that is centered at the origin and has a radius equal to 1. We will assume that angle measurements are always in radians, unless otherwise stated. By convention, an angle of θ radians is measured *counterclockwise* around the circle starting at the point $(1, 0)$, as shown in Figure 3.1.

The angle θ is always in standard position; one of the rays of the angle θ is *always* the x-axis. The terminal ray can lie in any quadrant. Suppose that the intersection of the unit circle with the terminal ray that forms the angle θ is the point P, whose coordinates are (x, y). We *define* $\sin \theta = y$ and $\cos \theta = x$. As θ and P move around the unit circle,

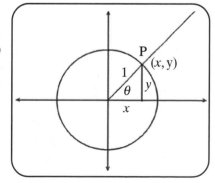

Figure 3.1.

the values of sin θ and cos θ will oscillate between 1 and −1. Negative angles are interpreted as the angle θ moving in the *clockwise* direction.

When the terminal ray of θ coincides with the y-axis, $x = 0$. From this, we see that when the terminal ray of θ coincides with the y-axis, cos $\theta = 0$. Similarly, when the terminal ray of θ coincides with the x-axis, $y = 0$ and sin $\theta = 0$. In other words, at the quadrantal angles, either cos $\theta = 0$ and sin $\theta = \pm 1$, or sin $\theta = 0$ and cos $\theta = \pm 1$. The values of the sine and cosine functions at the quadrantal angles will be very important in determining the domain of the other four trigonometric functions.

We can use the unit circle definitions of the sine and cosine functions to define the other four trigonometric functions:

$$\tan\theta = \frac{y}{x} \qquad \cot\theta = \frac{x}{y} \qquad \sec\theta = \frac{1}{x} \qquad \csc\theta = \frac{1}{y}$$

The coordinates of the point P will enable us to quickly determine the signs of the trigonometric functions. We can see that the sine of an angle will be positive if the y-coordinate of the point P is positive. In other words, if the point P lies in Quadrants I or II, then the y-coordinate of P will be *positive* and the sine of the angle formed will also be *positive*. If the point P lies in Quadrants III or IV, then the y-coordinate of P, and hence the sine of the angle formed, will be *negative*. Similarly, the cosine of the angle will be positive if the x-coordinate of the point P is positive. In other words, if the point P lies in Quadrants I or IV, the cosine of the angle formed will be positive, and if the point P lies in Quadrants II or III, the cosine of the angle formed will be negative. The signs of the sine and the cosine functions for the four quadrants are summarized in the table shown here.

Quadrant	Sign of the Sine	Sign of the Cosine
I	+	+
II	+	−
III	−	−
IV	−	+

For acute angles, the definitions of the sine and cosine functions using a right triangle are the same as the definitions based on the unit circle approach. For non-acute angles, the right triangle approach requires us to find a reference angle, whereas the unit circle approach does not require a reference angle. The point of intersection of the terminal ray and the unit circle is all that is needed to determine the sine and cosine of an angle using the unit circle approach.

With the right triangle approach to trigonometry, the sine and cosine of an angle are defined as the ratio of two sides of a right triangle. The unit circle approach allows us to easily interpret the sine and cosine of an angle as *functions*. The input, or independent variable, is the measure of an angle, and the output, or dependent variable, is one of the coordinates of the point where the terminal ray of the angle intersects the unit circle. Given any real number t, we can find a unique angle θ whose measure is t. As a result, the domain of the sine and cosine functions is the set of all real numbers. The output is either the x-coordinate or the y-coordinate of a point on the unit circle, and is therefore restricted to be a number between −1 and 1. The sine and cosine functions satisfy the following inequalities:

$$-1 \leq \sin \theta \leq 1 \qquad -1 \leq \cos \theta \leq 1$$
$$|\sin \theta| \leq 1 \qquad |\cos \theta| \leq 1$$

Because the sine and cosine functions are restricted to lie between −1 and 1, these functions are bounded. A function $f(x)$ is **bounded** if there is a finite number M that makes the inequality $|f(x)| \leq M$ true for all values of x. The sine and cosine functions are the *only* bounded trigonometric functions.

As we travel around the unit circle, we start to repeat ourselves. This repetition is referred to as the periodic behavior of the trigonometric functions.

The sine and cosine functions are **cyclic**, or periodic functions. A **periodic function** is a function that repeats itself after a certain, fixed amount has been added to the independent variable. This property can be represented algebraically as: $f(x + p) = f(x)$. The fixed constant p is called the period of the function. The **period** of a periodic function is the time needed for the function to complete one cycle. Using the unit circle as the basis for the definition of the sine and cosine functions, we can visualize the periodic nature of these functions. After going around the circle once, the values of the sine and cosine functions start to repeat. Each time the circle is traced out, 2π radians are added to the angle, so the period of the sine and cosine functions is 2π. The sine and cosine functions have the following property:

$$\sin (\theta + 2\pi n) = \sin \theta \text{ and } \cos (\theta + 2\pi n) = \cos \theta,$$
$$\text{where } n \text{ is any integer}$$

Because the period of $\sin \theta$ and $\cos \theta$ is 2π, once we know their values for all angles between 0 and 2π, their periodic nature will enable us to know their values for all possible angles.

From the definition of the tangent function, $\tan\theta = \frac{y}{x}$, the tangent function will have problems when the terminal ray of θ lies on the y-axis. At these angles, $x = 0$. The angles whose terminal ray lies on the y-axis are quadrantal angles whose measure is equal to an odd half-integer multiple of π: ..., $-\frac{3\pi}{2}, -\frac{\pi}{2}, \frac{\pi}{2}, \frac{3\pi}{2}, \frac{5\pi}{2}$.... We can also use the fact that $\tan\theta = \frac{\sin\theta}{\cos\theta}$ to determine the domain of the tangent function. The tangent function is defined as long as $\cos \theta \neq 0$, and the cosine of an angle is 0 when θ is a quadrantal angle whose measure is an odd half-integer multiple of π. The domain of the tangent function is the set of all real numbers such that $\cos \theta \neq 0$, or the set of real numbers excluding the odd half-integer multiples of π. The range of the tangent function is the set of all real numbers.

We can determine the domain of the other trigonometric functions in a similar manner. From the definition of the secant function, $\sec\theta = \frac{1}{x}$, the secant function will have the same problems that the tangent function had when $x = 0$. The domain of the secant function is the set of real numbers such that $\cos \theta \neq 0$, or the set of real numbers excluding the odd half-integer multiples of π. The secant function satisfies the inequality $|\sec \theta| \geq 1$.

From the definition of the cosecant function, $\csc\theta = \frac{1}{y}$, the cosecant function will be undefined when the terminal ray of θ lies on the x-axis. At these angles, $y = 0$. The angles whose terminal ray lies on the y-axis are quadrantal angles whose measure is equal to an integer multiple of π: ..., $-2\pi, -\pi, 0, \pi, 2\pi$.... We can also use the fact that $\csc\theta = \frac{1}{\sin\theta}$ to determine the domain of the cosecant function. The cosecant function is defined as long as $\sin \theta \neq 0$, and the sine of an angle is 0 when θ is a quadrantal angle whose measure is an integer multiple of π. The domain of the cosecant function is the set of all real numbers such that $\sin \theta \neq 0$, or the set of real numbers excluding the integer multiples of π. The cosecant function satisfies the inequality $|\csc \theta| \geq 1$.

From the definition of the cotangent function, $\cot\theta = \frac{x}{y}$, the cotangent function will have the same domain as the cosecant function. The domain of the cotangent function is the set of real numbers such that $\sin \theta \neq 0$, or the set of real numbers excluding the integer multiples of π. The range of the cotangent function is the set of all real numbers.

Symmetry is an important charac-
teristic of a function. Remember that
a function is even if its graph is symmetric with respect to the *y*-axis, and
the graph of an odd function is symmetric about the origin. Algebraically,
$f(x)$ is even if it satisfies the equation
$f(-x) = f(x)$, and $f(x)$ is odd if it satisfies the equation $f(-x) = -f(x)$. The
trigonometric functions are symmetric. Figure 3.2 shows the point *P* corresponding to an angle θ and the point

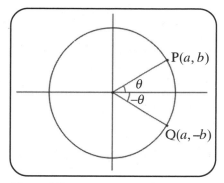

Figure 3.2.

Q corresponding to an angle $-\theta$. If the coordinates of *P* are (a, b), then
the coordinates for *Q* are $(a, -b)$. We have $\sin \theta = b$ and $\sin (-\theta) = -b$,
and in general, $\sin (-\theta) = -\sin \theta$; the sine function is an *odd* function.
Also, $\cos \theta = a$ and $\cos (-\theta) = a$, and in general, $\cos (-\theta) = \cos \theta$; the
cosine function is an *even* function.

The symmetry of the tangent function can be established from the
symmetry of the sine and cosine functions:

$$\tan(-\theta) = \frac{\sin(-\theta)}{\cos(-\theta)} = \frac{-\sin\theta}{\cos\theta} = -\tan\theta$$

The tangent function is an odd function, and its graph will be symmetric
with respect to the origin.

Lesson 3-2: Graphs of the Sine and Cosine Functions

In this lesson, we will explore the graphs of the trigonometric functions. We will begin with the graph of the sine function. Following in
the algebraic tradition, we will use *x* to represent the independent variable, or the angle input, and *y*, or $f(x)$, to represent the dependent variable: $y = f(x) = \sin x$. In Chapter 7, we will introduce a new coordinate
system: polar coordinates. The variables used to denote a graph in polar
coordinates are *r* and θ. To avoid any confusion, when we graph functions in the Cartesian coordinate system, we usually use *x* to represent
the independent variable, although it is also common to use *t* for the
independent variable when dealing with time. Keep in mind that all of
these variables are "dummy" variables. The particular symbol used to
represent a variable is not important, and in these matters it is convention that usually dictates the symbol choice.

Remember that the sine function is periodic, and its period is 2π. Because of this, it is only necessary to graph the sine function over one of its periods. Any interval of length 2π would serve; we could construct the entire graph of the sine function from a graph of the function over *any* interval of length 2π. For example, we can use the graph of the sine function over the interval $[0, 2\pi]$, or the interval $[\pi, 3\pi]$, or the interval $[1, 1 + 2\pi]$, or any other interval of length 2π, to construct the entire graph of the sine function. Remember also that the sine function is symmetric. It is an odd function, so its graph will be symmetric with respect to the origin. The symmetry of the sine function actually enables us to cut our work in half. If we know the graph of the sine function over the interval $[0, \pi]$, then because the sine function is an odd function, we can use its symmetry to determine its graph over the interval $[-\pi, 0]$. From this, we have the graph of the sine function over the interval $[-\pi, \pi]$, which is an interval of length 2π. Now we have the graph of the sine function over one of its periods, so we can graph the entire sine function by repeating the graph. This is one example of how a little thought can save a lot of work!

We will graph the function $y = \sin x$ over the interval $[0, \pi]$. From the symmetry of the sine function, we will obtain the graph of $y = \sin x$ over the interval $[-\pi, \pi]$. We will obtain the rest of the graph of the sine function by repeating this graph. Remember that the sine of an angle is the y-coordinate of the point where the terminal ray of the angle intersects the unit circle. The sine function is an odd function, so we know that it must pass through the origin: $\sin 0 = 0$. To visualize how the sine function changes as the angle x increases from 0 to π, we will use a unit circle, as shown in Figure 3.3.

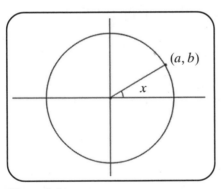

Figure 3.3.

As the angle, x, increases, the y-coordinate of the point of intersection also increases. Once x becomes the first quadrantal angle, $x = \frac{\pi}{2}$, the y-coordinate of the point of intersection attains its highest value: $\sin\frac{\pi}{2} = 1$. As the angle continues to increase, so that it now lies in Quadrant II, the y-coordinate of the point of intersection begins to decrease. Once the angle becomes the next quadrantal angle, $x = \pi$, the y-coordinate of the point of intersection has decreased back to 0: $\sin \pi = 0$. The graph of this part of the sine function is shown in Figure 3.4.

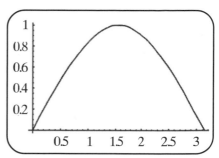

Figure 3.4.

Using the symmetry of the sine function, we can graph $y = \sin x$ over the interval $[-\pi, \pi]$. This graph is shown in Figure 3.5.

Using the periodic nature of the sine function, we can graph $y = \sin x$ by repeating this graph. The graph of $y = \sin x$ is shown in Figure 3.6.

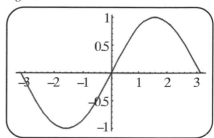

Figure 3.5.

There are some important properties of the sine function that should be noted. The domain of $y = \sin x$ is all real numbers, and the range is $[-1, 1]$. The sine function is an odd function, and its graph is symmetric with respect to the origin. The sine function is periodic, and its period is 2π. The sine function has a maximum value of 1, and it attains its maximum values at the points $x = \frac{\pi}{2} + 2\pi n$, where n is any integer. The sine function has a mini-mum value of -1, and it attains its

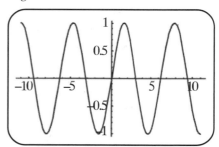

Figure 3.6.

minimum value at $x = -\frac{\pi}{2} + 2\pi n$, where n is any integer. The x-intercepts are integer multiples of π: $x = n\pi$.

The cosine function is also a periodic function and it is symmetric, and we can follow a similar approach in developing its graph. We will graph the function $y = \cos x$ over the interval $[0, \pi]$. From the symmetry of the cosine function, we will obtain the graph of $y = \cos x$ over the interval $[-\pi, \pi]$. We will obtain the rest of the graph of the sine function by repeating this graph. Remember that the cosine of an angle is the x-coordinate of the point where the terminal ray of the angle intersects the unit circle. The cosine function is an even function, so we know that its

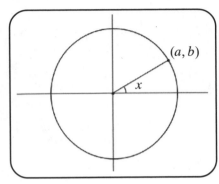

Figure 3.7.

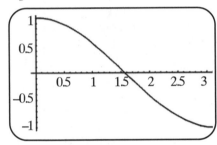

Figure 3.8.

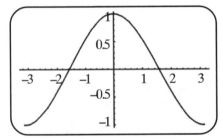

Figure 3.9.

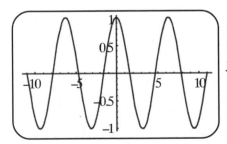

Figure 3.10.

graph will be symmetric with respect to the y-axis. To visualize how the sine function changes as the angle x increases from 0 to π, we will use a unit circle, as shown in Figure 3.7.

Because cos 0 = 1, the graph of the cosine function will start at the point (0, 1). The graph of the cosine function starts out at its maximum value. As the angle, x, increases, the x-coordinate of the point of intersection decreases. Once x becomes the first quadrantal angle, $x = \frac{\pi}{2}$, the x-coordinate of the point of intersection is 0: $\cos\frac{\pi}{2} = 0$. As the angle continues to increase, so that it now lies in Quadrant II, the x-coordinate of the point of intersection continues to decrease and become negative. Once the angle becomes the next quadrantal angle, x = π, the x-coordinate of the point of intersection takes on its minimum value: cos π = −1. The graph of this part of the cosine function is shown in Figure 3.8.

Using the symmetry of the cosine function, we can graph y = cos x over the interval [−π, π]. This graph is shown in Figure 3.9.

Using the periodic nature of the cosine function, we can graph y = cos x by repeating this graph. The graph of y = cos x is shown in Figure 3.10.

There are some important properties of the cosine function that should be noted. The domain of y = cos x is all real numbers, and the range is [−1, 1]. The cosine function is an even function, and its graph is symmetric with respect to the *y*-axis. The cosine function is periodic, and its period is 2π. The cosine function has a maximum value of 1, and it attains its maximum values at the points $x = 0 + 2\pi n$, or $x = 2\pi n$, where *n* is any integer. The collection of points $2\pi n$, where *n* is an integer, is called the **even integer multiples** of π. The cosine function has a minimum value of -1, and it attains its minimum value at $x = \pi + 2\pi n$, or $x = (2n + 1)\pi$, where *n* is any integer. The collection of points $(2n + 1)\pi$, where *n* is any integer, is called the **odd integer multiples** of π. The *x*-intercepts are located at the **odd half-integer multiples** of π: $x = \dfrac{(2n+1)\pi}{2}$.

Lesson 3-3: Graphing the Other Trigonometric Functions

The secant and cosecant functions are the reciprocals of the cosine and sine functions, respectively. The graphs of the secant and cosecant function are directly related to the graphs of the cosine and sine functions. We will first analyze the graph of the secant function.

The secant function is a periodic function, and its period is 2π. The secant function inherits the symmetry of the cosine function: the secant function is an even function. The fact that $\left| \cos x \right| \le 1$ means that $\left| \sec x \right| \ge 1$. Because the *magnitude* of the cosine function is 1 when *x* is an integer multiple of π ($\left| \cos n\pi \right| = 1$), we know that the *magnitude* of the secant function is also 1 when *x* is an integer multiple of π ($\left| \sec nx \right| = 1$).

We also know that $\cos\left(\dfrac{(2n+1)\pi}{2}\right) = 0$,

which means that the secant function will have vertical asymptotes at $x = \dfrac{(2n+1)\pi}{2}$. To graph the secant func-

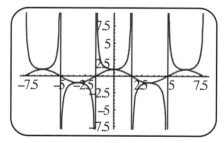

tion, start with the graph of the co-sine function. The *x*-intercepts of the   *Figure 3.11.*
cosine function will be vertical asymptotes of the secant function. The two graphs will intersect when *x* is an integer multiple of π. The graphs of $y = \cos x$ and $y = \sec x$ are shown in Figure 3.11.

We will graph the cosecant using the same approach. The cosecant function is a periodic function, and its period is 2π. The cosecant function inherits the symmetry of the sine function: the cosecant function is an odd function. The fact that $|\sin x| \le 1$ means that $|\csc x| \ge 1$. Because the *magnitude* of the sine function is 1 when x is an odd half-integer multiple of π, ($\left|\sin\left(\frac{(2n+1)\pi}{2}\right)\right| = 1$), we know that the *magnitude* of the cosecant function is also 1 when x is an odd half-integer multiple of π ($\left|\csc\left(\frac{(2n+1)\pi}{2}\right)\right| = 1$). We also know that $\sin n\pi = 0$, which means that the

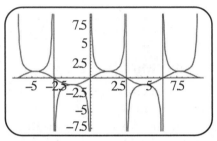

Figure 3.12.

cosecant function will have vertical asymptotes at $x = n\pi$. To graph the cosecant function, start with the graph of the sine function. The x-intercepts of the sine function will be vertical asymptotes of the cosecant function. The two graphs will intersect when x is an odd half-integer multiple of π. The graphs of $y = \sin x$ and $y = \csc x$ are shown in Figure 3.12.

The graph of the tangent function will require a little more work. The tangent function is periodic, but its period is π, *not* 2π, as is the case with the sine and cosine functions. This can be established by analyzing $\tan(\theta + \pi)$ for angles θ that lie in each of the four quadrants. If θ is an angle that lies in Quadrant I or Quadrant III, then adding π to θ will move the angle from Quadrant I to Quadrant III, or from Quadrant III to Quadrant I, but adding π will *not* change the reference angle. The tangent function is positive in both of these quadrants, and, as a result, $\tan(\theta + \pi) = \tan \theta$. If, on the other hand, θ is an angle that lies in Quadrant II or Quadrant IV, then adding π to θ will move the angle from Quadrant II to Quadrant IV, or from Quadrant IV to Quadrant II, but adding π will not change the reference angle. The tangent function is negative in both of these quadrants, and, as a result, $\tan(\theta + \pi) = \tan \theta$. We will learn some trigonometric identities in Chapter 5 that will enable us to prove this algebraically. For now, remember that the period of the tangent function (and, by association, the cotangent function) is π.

In order to graph the tangent function, we only need to analyze the function over an interval of length π. We could use any interval of length π. If our interval is symmetric, then we can use the symmetry of the tangent function to cut down on our work. Remember that the tangent function is an odd function, so its graph will be symmetric with respect to the origin. If we analyze the tangent function over the interval $\left[0,\frac{\pi}{2}\right]$, then by symmetry we will have the graph of the tangent function over the interval $\left[-\frac{\pi}{2},\frac{\pi}{2}\right]$. The periodic nature of the tangent function can be used to graph the tangent function over any interval.

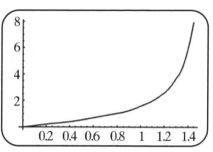
Figure 3.13.

Remember that $\tan x = \frac{\sin x}{\cos x}$. The tangent function, being an odd function, must pass through the origin. As x increases from 0 to $\frac{\pi}{2}$, $\sin x$ increases from 0 to 1 and $\cos x$ decreases from 1 to 0. If the numerator of a ratio increases while its denominator decreases, the net effect will be that the ratio increases. Because $\cos \frac{\pi}{2} = 0$, the tangent function will have a vertical asymptote at $x = \frac{\pi}{2}$. From this we see that as x increases from 0 to $\frac{\pi}{2}$, $\tan x$ will increase from 0 to ∞. The graph of $y = \tan x$ over the interval $\left[0,\frac{\pi}{2}\right]$ is shown in Figure 3.13.

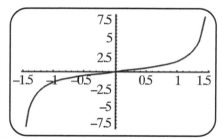
Figure 3.14.

Using the symmetry of the tangent function, we can graph $y = \tan x$ over the interval $\left[-\frac{\pi}{2},\frac{\pi}{2}\right]$. This graph is shown in Figure 3.14.

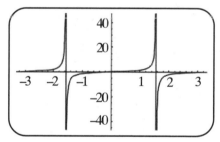
Figure 3.15.

Using the period nature of the tangent function, we can graph $y = \tan x$ over several periods. The graph of $y = \tan x$ is shown in Figure 3.15.

The graph of the cotangent function is the reciprocal of the graph of the tangent function, and is shown in Figure 3.16.

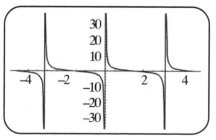
Figure 3.16.

Lesson 3-4: Transformations

Now that we know how to graph the six trigonometric functions, we can move them around the coordinate plane. Trigonometric functions can be shifted (vertically and horizontally), stretched/contracted (vertically or horizontally), and reflected (about the x-axis or the y-axis). Before we examine these transformations in more detail, we need to define some terminology.

We have already discussed the period of the trigonometric functions. In general, the period of a function is the time it takes to go through one cycle. The period of the sine and cosine functions is 2π, and the period of the tangent function is π. We will denote the period of a trigonometric function by T. In Chapter 1, we discussed the process of stretching or contracting the graph of a function. We can apply that knowledge to transform the graphs of the trigonometric functions. If we stretch or contract the functions *horizontally*, we will change the *period* of the functions. Remember that a horizontal stretch or contraction results when you multiply the *argument* of a function by a constant. When we transformed functions in Chapter 1, we used c to represent the constant that we used to shift or stretch a function. With trigonometric functions, it is customary to turn to the Greek alphabet to find a symbol to represent this constant, and the letter of choice is omega, ω. We will use ω to represent the factor by which we stretch or shrink the graph of a trigonometric function.

The graph of the function $\sin(\omega x)$ (or $\cos(\omega x)$) is the graph of $\sin x$ (or $\cos x$) horizontally stretched or contracted by a factor of $\frac{1}{|\omega|}$. One cycle of the sine or cosine function is normally completed over an interval of length 2π. If the function is stretched or contracted by a factor of $\frac{1}{|\omega|}$, then so is the length required to complete one cycle. Thus the period changes from 2π to $\frac{2\pi}{|\omega|}$. The graphs of $y = \sin x$ and $y = \sin 3x$ (or $\sin(\omega x)$ where $\omega = 3$) are shown in Figure 3.17.

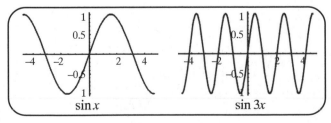

sin x   sin 3x

Figure 3.17.

In general, the functions $y = \sin(\omega x)$ and $y = \cos(\omega x)$ will be periodic functions, and their period is given by the formula:

$$T = \frac{2\pi}{|\omega|}$$

The period of the function $y = \tan(\omega x)$ is given by the formula:

$$T = \frac{\pi}{|\omega|}$$

The **amplitude** of a bounded periodic function is one-half the distance between the maximum and minimum values of the function. Because the sine and cosine functions oscillate between 1 and -1, the amplitude of the sine and cosine functions is 1: $\frac{1-(-1)}{2}$. In general, the functions $y = A \sin x$ and $y = A \cos x$ will satisfy the inequalities:

$$-|A| \le A \sin x \le |A| \text{ and } -|A| \le A \cos x \le |A|$$

The constant $|A|$ is called the *amplitude* of the function. The amplitude of a function is always positive. The tangent, cotangent, secant, and cosecant functions do not have an amplitude. These functions are unbounded, so the maximum and minimum values of these functions are meaningless.

The **average value** of the sine and cosine functions is the average of the maximum and minimum values of the function. The maximum value of the sine and cosine functions is 1, and the minimum value is -1. The average value of the sine and cosine

functions is 0: $\frac{1+(-1)}{2} = 0$.

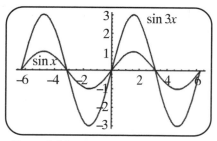

Figure 3.18..

If we stretch or contract the sine or cosine functions *vertically*, we will change the amplitude of these functions. Remember that a vertical stretch or contraction results when you multiply the function by a constant. Stretching or contracting the sine or cosine functions vertically will not change their *average value* because the maximum and minimum values will both be changed by the same factor and their average will still be 0. The graph of $y = 3 \sin x$ is the graph of $y = \sin x$ stretched vertically by a factor of 3. The maximum value of $y = 3 \sin x$ is 3, and the minimum value is –3. The amplitude of $y = 3 \sin x$ is $A = \frac{3-(-3)}{2} = 3$, and its average value is 0. The graphs of $y = 3 \sin x$ and $y = \sin x$ are shown in Figure 3.18.

Example 1

Find the amplitude and the period of the following functions:

a. $y = 2 \sin 3x$　　　　b. $y = -3 \cos 5x$　　　c. $y = 8 \tan 4x$

Solution: Use the formulas for the period and the amplitude of the trigonometric functions:

a. $y = 2 \sin 3x$: The amplitude is 2, and the period is $T = \frac{2\pi}{3}$.

b. $y = -3 \cos 5x$: The amplitude is 3, and the period is $T = \frac{2\pi}{5}$.

c. $y = 8 \tan 4x$: It makes no sense to talk about the amplitude of

the tangent function, and the period is $T = \frac{\pi}{4}$.

We can also translate the graphs of these functions vertically or horizontally. When we translate these graphs, we will not change the period or the amplitude of these functions. We may, however, change the intercepts and the location of any vertical asymptotes. A vertical translation will change the average value of the sine and cosine functions. The average value of the vertically translated function will be the amount of the vertical shift. The graph of the function $y = \sin x + 5$ is the graph of $y = \sin x$ shifted up 5 units. The amplitude of $y = \sin x + 5$ is 1 and its period is 2π. The graphs of $y = \sin x$ and $y = \sin x + 5$ are shown in Figure 3.19. Notice that the function $y = \sin x + 5$ has no x-intercepts. The function $y = \sin x + 5$ satisfies the inequality $4 \le \sin x + 5 \le 6$. In other words, its maximum value is 6 and its minimum value is 4. From this, we see that its average value is 5: $\frac{6+4}{2} = 5$, which is exactly the value of the

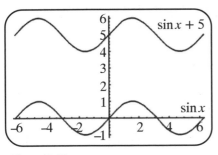

Figure 3.19.

vertical shift. It is important to keep in mind that the sine and cosine functions are bounded between −1 and 1. There are times when that fact will be very useful.

Example 2

Sketch the graphs of the following functions:

a. $y = 3 \cos 2x$ b. $y = 2 \sin x + 4$

Solution: Analyze each graph according to the transformations involved.

a. $y = 3 \cos 2x$: The cosine function has been compressed

horizontally by a factor of 2, so that its period is $T = \frac{2\pi}{2} = \pi$.

The amplitude of $y = 3 \cos 2x$ is 3. The graph of the cosine function will be contracted horizontally by a factor of 2 and stretched vertically by a factor of 3. The graphs of $y = 3 \cos 2x$ and $y = \cos x$ are shown in Figure 3.20.

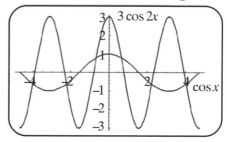

Figure 3.20.

b. $y = 2 \sin x + 4$: The sine function has been stretched vertically by a factor of 2, so its amplitude is 2. The graph of the sine function will be shifted up 4 units. Using the order of operations, the sine function must be stretched vertically by a factor of 2 and then shifted up by 4 units. The graphs of $y = 2 \sin x + 4$ and $y = \sin x$ are shown in Figure 3.21.

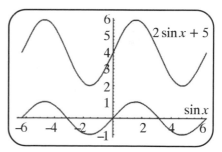

Figure 3.21.

We can also reflect the graph of a trigonometric function across the x-axis and across the y-axis. To reflect the graph of a function across the x-axis, multiply the function by −1. The graph of the function $y = -2 \cos x$

is the graph of the cosine function vertically stretched by a factor of 2 and then reflected across the x-axis, as shown in Figure 3.22.

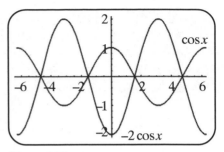

Figure 3.22.

To reflect the graph of a function across the y-axis, multiply the *argument* of the function by −1. The graph of the function $y = 3 \sin (-x)$ is the graph of the sine function vertically stretched by a factor of 3 and then reflected across the y-axis, as shown in Figure 3.23. Because of the symmetry of the sine function, we have that:

$$y = 3 \sin (-x) = -3 \sin x.$$

From this relationship, we can see that the effect of reflecting the graph of the sine function across the y-axis is the same as reflecting its graph across the x-axis. There's more than one way to look at the transformation of a trigonometric function.

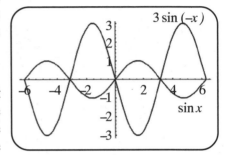

Figure 3.23.

When we translate a trigonometric function horizontally, we introduce what is called a phase shift. We will discuss this transformation in more detail in the next lesson.

Lesson 3-4 Review

1. Find the period and the amplitude of $y = -2 \sin 5x + 8$.

2. Find the period and sketch the graph of $y = \tan 2x + 1$.

Lesson 3-5: Phase Shift

The graphs of the sine and cosine curves are called *sinusoidal* curves. A **sinusoidal** curve is a curve that has the shape of a sine curve. The graph of the cosine function is also considered to be a sinusoidal curve. We can stretch or contract the graphs of the trigonometric functions by multiplying the *function* by a constant (for a vertical stretch) or by multiplying the *argument* by a constant (for a horizontal stretch). Vertical stretches will change the amplitude of the function, and horizontal stretches will change the period of the function. Adding a constant to the function results in a vertical shift, and adding a constant to the argument of the function will result in a horizontal shift. A horizontal shift is called a *phase shift*.

A sinusoidal graph has the general form:

$$y = A\sin(\omega x - \phi) = A\sin\left[\omega\left(x - \frac{\phi}{\omega}\right)\right]$$

where ω and ϕ are real numbers with $\omega > 0$. The graph of this function will be a sine curve with amplitude $|A|$ and period $T = \frac{2\pi}{\omega}$. To understand this function better, let us examine $A\sin(\omega x - \phi)$ in more detail. The sine function will complete one cycle when its argument varies from 0 to 2π. The function will start out when the argument is equal to 0:

$$\omega x - \phi = 0, \text{ or } x = \frac{\phi}{\omega}$$

After one complete cycle, the argument is equal to 2π:

$$\omega x - \phi = 2\pi, \text{ or } x = \frac{\phi}{\omega} + \frac{2\pi}{\omega}$$

From this, we see that the period is $T = \frac{2\pi}{\omega}$, and the graph of $A\sin(\omega x - \phi)$ starts out at $\frac{\phi}{\omega}$ instead of 0. In other words, the graph of $y = A\sin\left[\omega\left(x - \frac{\phi}{\omega}\right)\right]$ will be the graph of $A\sin\omega x$ shifted horizontally by $\frac{\phi}{\omega}$ units. The number $\frac{\phi}{\omega}$ is called the **phase shift** of the graph.

Example 1

Find the amplitude, the period, and the phase shift of

$$y = 4\sin\left(3x - \frac{\pi}{2}\right).$$

Solution: The amplitude is 4, the period is $T = \frac{2\pi}{3}$, and the

phase shift is $\frac{\frac{\pi}{2}}{3} = \frac{\pi}{6}$.

The following strategy will enable you to sketch the graph of a sinusoidal function of the form $y = A \sin(\omega x - \phi)$ or $y = A \cos(\omega x - \phi)$.

1. Determine the amplitude, $|A|$, the period, $T = \frac{2\pi}{\omega}$, and phase shift, $\frac{\phi}{\omega}$, of the function.

2. Determine the starting point for one cycle of the graph by setting the argument of the function equal to 0: $x = \frac{\phi}{\omega}$. Determine the ending point for that cycle by setting the argument of the function equal to 2π: $x = \frac{\phi}{\omega} + \frac{2\pi}{\omega}$.

3. Evaluate and graph the function at five values of x. To find these key values, start with the starting point of the interval and increase x in increments of $\frac{T}{4} = \frac{\pi}{2\omega}$: $x = \frac{\phi}{\omega}$, $x = \frac{\phi}{\omega} + \frac{\pi}{2\omega}$, $x = \frac{\phi}{\omega} + \frac{\pi}{\omega}$, $x = \frac{\phi}{\omega} + \frac{3\pi}{2\omega}$, and $x = \frac{\phi}{\omega} + \frac{2\pi}{\omega}$.

4. Draw a sinusoidal graph that passes through those five points. That is the graph of one cycle of the function. Extend the graph in both directions to obtain the complete graph.

Example 2

Sketch a graph of the function $y = 4\cos\left(3x + \frac{\pi}{2}\right)$.

Solution: Follow the strategy outlined previously:

1. The amplitude is 4, the period is $T = \frac{2\pi}{3}$, and the phase shift is $\dfrac{-\frac{\pi}{2}}{3} = -\frac{\pi}{6}$.

2. The starting point is $x = -\frac{\pi}{6}$, and the stopping point is $x = -\frac{\pi}{6} + \frac{2\pi}{3} = \frac{\pi}{2}$.

3. Evaluate the function at the five key values. To find the key values, increase x in increments of $\frac{T}{4} = \dfrac{\frac{2\pi}{3}}{4} = \frac{\pi}{6}$:

x	$y = 4\cos\left(3x + \frac{\pi}{2}\right)$
$x = -\frac{\pi}{6}$	4
$x = 0$	0
$x = \frac{\pi}{6}$	−4
$x = \frac{2\pi}{6}$	0
$x = \frac{3\pi}{6} = \frac{\pi}{2}$	4

4. Connect the dots to graph one period. Then extend your graph in both directions to get the complete graph.

The graphs of $y = 4\cos\left(3x + \frac{\pi}{2}\right)$ and $y = \cos x$ are shown in Figure 3.24.

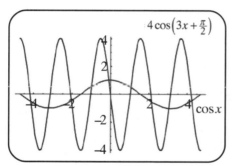

Figure 3.24.

Lesson 3-5 Review

1. Find the amplitude, the period, and the phase shift of
$y = 4\cos\left(2x - \frac{\pi}{5}\right)$.

2. Sketch a graph the function $y = -3\sin\left(4x - \frac{2\pi}{3}\right)$.

Lesson 3-6: Applications

The sine and cosine functions are particularly useful in describing physical phenomena that are cyclic in nature. For example, average monthly temperature, the number of hours of daylight, tide height, and current through a circuit are all examples of cyclic behavior that can be described using a sinusoidal function. In this lesson, we will learn how to construct sinusoidal models to describe oscillating behavior.

· The general form of a sinusoidal function is:

$$y = A \sin\left[\omega\left(x - \frac{\phi}{\omega}\right)\right] + B$$

where $|A|$ is the amplitude, $\frac{2\pi}{\omega}$ is the period, $\frac{\phi}{\omega}$ is the phase shift, and the constant B indicates the vertical shift. From the description of oscillating behavior, we can find each of these quantities. When creating a model for oscillating behavior, it does not matter whether you use a sine function or a cosine function. Sine functions are usually used because the sine function is 0 at 0, and this fact helps solve for various constants a little more easily. Keep in mind that it is not wrong to use a cosine function. The constants will be different, but the two formulas will be equivalent.

Example 1

The voltage, V, of an electrical outlet in a home is given as a function of time, t, by the formula $V = 120 \cos(120\pi t)$, where V is measured in volts and t is measured in seconds. Find the amplitude and the period of oscillation.

Solution: The amplitude is 120 volts, and the period is given by:

$$T = \frac{2\pi}{120\pi} = \frac{1}{60}.$$

From the period, we see that each cycle takes $\frac{1}{60}$ sec, or that there are 60 cycles per second. The **frequency** of an oscillation is the number of cycles per second. One unit for frequency is **hertz**, and is denoted by the symbol hz:

1 hz = 1 cycle per second

A 120-volt household outlet produces current at 60 hz.

Example 2

In Anchorage, Alaska, the shortest day of the year has 5 hours 26 minutes of daylight and occurs on December 21st. The longest day of the year has 19 hours 22 minutes of daylight and occurs on June 21st. Find a function that models the number of daylight hours, H, as a function of the number of days since January 1st, d.

Solution: We need to find a function of the form

$H = A \sin[\omega(d - \phi)] + B$.

The amplitude, A, is one-half the difference between the maximum and minimum hours of daylight:

$$A = \frac{19\frac{22}{60} - 5\frac{26}{60}}{2} = \frac{13\frac{56}{60}}{2} = 6\frac{58}{60} \approx 6.97$$

There are 365 days in a year, so the period is 365 days. From this we can solve for ω:

$$T = \frac{2\pi}{\omega}$$

$$\omega = \frac{2\pi}{T} = \frac{2\pi}{365}$$

Now, the average number of daylight hours is:

$$\tfrac{1}{2}\left(19\tfrac{22}{60} + 5\tfrac{26}{60}\right) = \tfrac{1}{2}\left(24\tfrac{48}{60}\right) = 12\tfrac{24}{60} = 12.4$$

Remember that the average value of the sine function is the amount of the vertical shift of the translated function, so we know that the graph of the sine function will be shifted upward by 12.4 units: $B = 12.4$. On June 21st, the curve will attain its maximum value. This means that when $d = 172$, $\omega(d - \phi) = \frac{\pi}{2}$.

We know ω and we know d, so we can solve for ϕ:

$$\omega(d - \phi) = \frac{\pi}{2}$$

Substitute in for ω and d: $\dfrac{2\pi}{365}(172 - \phi) = \dfrac{\pi}{2}$

Multiply both sides of the equation by $\frac{365}{2\pi}$:

$$(172-\phi)=\frac{365}{4}$$

$$\phi=172-\frac{365}{4}=\frac{323}{4}=80.75$$

Our model for the number of daylight hours is:

$$H=6.97\sin\left[\frac{2\pi}{365}(d-80.75)\right]+12.4$$

Lesson 3-6 Review

1. Suppose that high tide occurred at 3:38 a.m. and low tide occurred at 10:08 a.m. The height of the water at high tide was 8.4 feet and the height of the water at low tide was –1.2 feet. Find a sinusoidal function of the form $H=A\sin[\omega(t-\phi)]+B$ that can be used to model the height of the water, H, as a function of time since midnight, t.

Answer Key
Lesson 3-4 Review

1. The period is $T=\frac{2\pi}{5}$, and the amplitude is 2.

2. The period is $T=\frac{\pi}{2}$, and the graph of $y=\tan 2x+1$ is shown in Figure 3.25.

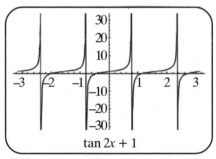

Figure 3.25.

Lesson 3-5 Review

1. The amplitude is 4, the period is $T = \frac{2\pi}{2} = \pi$, and the phase shift is $\frac{\pi}{10}$.

2. The amplitude is 3, the period is $T = \frac{2\pi}{4} = \frac{\pi}{2}$, and the phase shift is $\frac{\pi}{6}$.

The graph of $y = -3\sin\left(4x - \frac{2\pi}{3}\right)$ is shown in Figure 3.26.

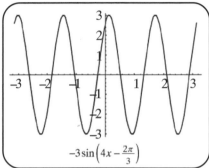

Figure 3.26.

Lesson 3-6 Review

1. The amplitude is 4.8, and the average value (and hence the vertical shift) is 3.6. The time between the maximum and minimum values is 6 hours and 30 minutes, or 6.5 hours. This time is one half of a period, which means that the period is 13 hours.

Use the equation $T = \frac{2\pi}{\omega}$ to solve for ω: $13 = \frac{2\pi}{\omega}$, or $\omega = \frac{2\pi}{13}$.

At this point, our model is the function $H = 4.8\sin\left[\frac{2\pi}{13}(t-\phi)\right] + 3.6$.

To find the phase shift, set the argument of the sine function equal to $\frac{\pi}{2}$.

At this time, the tide is high, which means $t = 3\frac{19}{30}$.

We can then solve for ϕ:

$$\frac{2\pi}{13}\left(3\frac{19}{30} - \phi\right) = \frac{\pi}{2}$$

$$\left(3\frac{19}{30} - \phi\right) = \frac{13}{4}$$

$$\phi = 3\frac{19}{30} - \frac{13}{4} = \frac{23}{60}$$

Our model is $H = 4.8\sin\left[\frac{2\pi}{13}\left(t - \frac{23}{60}\right)\right] + 3.6$.

Inverse Trigonometric Functions

A function is used to represent the relationship, or dependence, between one quantity and another. The rate at which a cricket chirps is related to, or is a function of, the temperature of its environment. Under certain circumstances, we can use a cricket as a thermometer. In this situation, we are looking at the environmental temperature as a function of the chirp rate. Looking at things from the reverse perspective can be thought of mathematically as inverting a function. The focus of this chapter is to invert the six basic trigonometric functions, and to examine the relationships between these inverse functions.

Lesson 4-1: Inverse Functions

There are rules for how to invert a function, and there are conditions under which a function *cannot* be inverted. We denote the inverse of the function f by f^{-1}. When dealing with rational and exponential expressions, we have interpreted the superscript -1 to denote the reciprocal of a number. Now we are using it to denote the inverse of a function. There is a reason that we are reusing this notation. The *reciprocal* of a number a is the number to *multiply a* by to get 1 (the *multiplicative identity*). The **inverse** of a function f is a function that you *compose* with f to get the *identity function*.

An identity is something that does not change the input. For example, the *additive* identity is the number that does not change any number under addition: the additive identity is 0. Similarly, the *multiplicative* identity is the number that does not change any number under multiplication: the multiplicative identity is 1. The **identity function** is the function that does not change any input value of the function: the identity function is $f(x) = x$. Notice that the input of the function is the same as the output of the function. The inverse of a function f is the unique function, denoted f^{-1}, that has the following relationship to f:

$$\left(f^{-1} \circ f\right)(x) = x \text{ and } \left(f \circ f^{-1}\right)(x) = x$$

If a function has an inverse, the function is **invertible**, and the inverse of an invertible function is unique. The statement $y = f^{-1}(x)$ means the same thing as $f(y) = x$. We will use this idea repeatedly when we are working with inverse functions.

As I mentioned earlier, the inverse of a function involves a change in perspective: the output becomes the input, and the input becomes the output. If f is an invertible function with domain X and range Y, then f^{-1} will be a function with domain Y and range X. When inverting a function, the role of the independent variable, x, and the role of the dependent variable, y or f, switch. This concept will be useful in actually finding the inverse of invertible functions. If an invertible function f passes through the point (a, b), then its inverse, f^{-1}, will pass through the point (b, a): the input becomes the output, and the output becomes the input. If f represents the chirp rate as a function of temperature, where temperature is measured in degrees Fahrenheit, and $f(60) = 100$, then this means that at a temperature of $60°$ F, the chirp rate is 100 chirps per minute. From this, we know that $f^{-1}(100) = 60$, meaning that a chirp rate of 100 chirps per minute corresponds to an environmental temperature of $60°$ F.

If a function f takes x to y, then the inverse function, f^{-1}, takes y back to x. From another perspective, f^{-1} undoes what f does: if we start with x and apply f, and then we apply f^{-1}, the final result will be x. If we apply a function and then we apply its inverse, we will arrive exactly where we started. Also, f undoes what f^{-1} does.

Working with inverses can get a bit tricky. The notation may be awkward at first, but with some practice it should become easier. It is important to keep in mind that if a function f is invertible and $f(a) = b$, then $f^{-1}(b) = a$. In other words, if f is invertible, then $f^{-1}(r) = s$ means the same thing as $f(s) = r$.

Not all functions have inverses. A function must be one-to-one in order for it to have an inverse. Our understanding of symmetry allows us to conclude that an even function is not invertible on its entire domain.

Remember that an even function satisfies the relationship $f(-x) = f(x)$, so if $f(a) = b$, then $f(-a) = b$ and there is no unique choice for $f^{-1}(b)$. On the other hand, every linear function is invertible.

A function will be invertible if it is one-to-one. A function is one-to-one if no two elements in the domain have the same image. In other words, if a and b are in the domain of f, and $a \neq b$, then $f(a) \neq f(b)$. Another way to look at this situation is that if $f(a)$ and $f(b)$ are two points in the range with $f(a) = f(b)$, then $a = b$. Determining whether a function has an inverse is equivalent to determining whether the function is one-to-one.

As we discussed in Chapter 1, the function $f(x) = x^2$ is not one-to-one, and so it is not invertible. If a function is not one-to-one on its domain, we may be able to restrict the domain so that it is one-to-one on its restricted domain, which would then *make* the function invertible (on that restricted domain).

If a function is both increasing and decreasing on an interval I, then it cannot be one-to-one on that interval. If a function is not one-to-one on an interval, then it cannot have an inverse on that interval. The domain of the function would have to be restricted so that only one type of behavior, either increasing or decreasing, is exhibited.

If an invertible function is defined by a formula, it is sometimes possible to find a formula for the inverse function. The key to finding a formula for the inverse of a function is to switch the role of the independent variable and the role of the dependent variable. In other words, the role of x and the role of y switch when you invert a function. To find the inverse of a function $y = f(x)$, first switch x and y, and then solve for y. To switch x and y, everywhere you see an x, write a y, and everywhere you see a y, replace it with an x.

Example 1

Find the inverse of the function $f(x) = 2x + 1$.

Solution: The function $f(x) = 2x + 1$ is a linear function, so it will have an inverse. Rewrite the function as $y = 2x + 1$. Switch x and y and then solve for y: $y = 2x + 1$ becomes $x = 2y + 1$. Now we can solve for y:

$x = 2y + 1$

$2y = x - 1$

$$y = \frac{x-1}{2}$$

Thus $f^{-1}(x) = \frac{1}{2}(x-1)$.

Notice that the function $f(x) = 2x + 1$ takes the input, doubles it, and then adds 1. The inverse function, f^{-1}, will just undo what the function f does, but in reverse order. First, f^{-1} undoes the addition of 1 by subtracting 1. Then f^{-1} undoes the multiplication by 2, by dividing by 2. You can see this in the formula for f^{-1}: $f^{-1}(x) = \frac{1}{2}(x-1)$.

Being able to actually solve for the inverse of a function can be difficult, and may require the definition of a new function (as in the case with exponential functions), but if you have the graph of an invertible function, there is a way to graph its inverse.

In general, if a function passes through the point (a, b), then its inverse will pass through the point (b, a). The only points that a function has in common with its inverse will be the points that do not change when x and y are switched. These are the points where $y = x$, or $f(x) = x$. Consider the function $f(x) = x^3$ and its inverse, $f^{-1}(x) = \sqrt[3]{x}$. The graphs of these two functions are shown in Figure 4.1. These graphs are the reflections of each other about the line $y = x$. In general, to graph the inverse of a function f, start with the graph of f and reflect the graph about the line $y = x$.

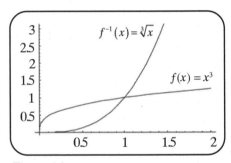

Figure 4.1.

When inverting a function, the roles of x and y switch. As a result, if an invertible function has a vertical asymptote $x = a$, then its inverse function will have a horizontal asymptote $y = a$. Similarly, if an invertible function has a horizontal asymptote $y = b$, then its inverse function will have a vertical asymptote $x = b$.

Lesson 4-1 Review

1. Find the inverse of the function $f(x) = 3x - 4$.

Lesson 4-2: The Inverse Sine and Cosine Functions

Only functions that are one-to-one have inverses. This is equivalent to requiring a function to be monotonic. Periodic functions can have specific regions where they are increasing or decreasing, but, by their very nature, periodic functions cannot be monotonic on their entire domain.

The reason that a periodic function cannot be monotonic on its entire domain is that a periodic function with period p must satisfy the relationship $f(x + p) = f(x)$. From this, we see that $x < x + p$, yet $f(x)$ is *not* less than $f(x + p)$, so a periodic function cannot be an increasing function on its domain. Similarly, a periodic function cannot be a decreasing function on its domain. Looking at it from another perspective, a periodic function has to repeat, so if a periodic function starts off increasing, it will have to decrease in value somewhere along the line in order for the function to return to its original value and begin a new cycle.

The periodic nature of the trigonometric functions means that the trigonometric functions are not one-to-one. Functions that are not one-to-one cannot have an inverse, but we *can* restrict the domain of the trigonometric functions so that they *are* one-to-one, and hence invertible, on their restricted domain.

When finding the inverse of a function, it is helpful to keep in mind what the domain and the range of the function represent. This will help clarify some of the properties of inverse functions. The triangle definitions of the trigonometric functions will be helpful in this analysis. With the function $f(\theta) = \sin \theta$, θ is the measure of an angle, and $f(\theta)$ is a ratio of two lengths, and is therefore dimensionless. Also, the values that $f(\theta)$ take on must lie between -1 and 1. If we restrict the domain of f so that it is a one-to-one function, then the expression f^{-1} will be a function whose independent variable (or domain) consists of real numbers between -1 and 1, and whose dependent variable (or range) consists of angle measures. We write the inverse sine function as either $\sin^{-1} x$ or $\arcsin x$. The restricted domain for the $f(x) = \sin x$ is $-\frac{\pi}{2} \le x \le \frac{\pi}{2}$, and the range is $-1 \le \sin x \le 1$. The arcsine function is the inverse of the sine function over this restricted domain. The domain of the arcsine function is $-1 \le x \le 1$ and the range is $-\frac{\pi}{2} \le \arcsin x \le \frac{\pi}{2}$. The equation $y = \arcsin x$ *means* $x = \sin y$. The graph of the arcsine function and the graph of the (restricted) sine function reflected across the line $y = x$ is shown in Figure 4.2 on page 82.

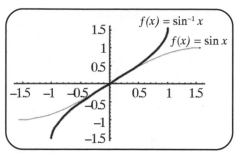

Figure 4.2.

The graph of the arcsine function is symmetric about the origin. The arcsine function is an odd function: $\sin^{-1}(-x) = -\sin^{-1}x$. The arcsine function inherits the symmetry of the sine function because of the symmetry of the domain and range of the arcsine function.

Example 1

Find exact values of the following:

a. $\sin^{-1}1$

b. $\sin^{-1}\left(-\frac{1}{2}\right)$

Solution:

a. There are infinitely many angles whose sine is 1, but we are looking for the unique angle that lies in the restricted domain of the arcsine function: $-\frac{\pi}{2} \le \arcsin x \le \frac{\pi}{2}$:

$\sin^{-1}1 = \frac{\pi}{2}$.

b. The range of the arcsine function corresponds to angles that lie in either Quadrant I (where the sine is positive) or Quadrant IV (where the sine is negative). To find $\sin^{-1}\left(-\frac{1}{2}\right)$, we need to first find the reference angle whose sine is $\frac{1}{2}$: $\sin^{-1}\frac{1}{2} = \frac{\pi}{6}$. Next, put the angle in the correct quadrant: $\sin^{-1}\left(-\frac{1}{2}\right) = -\frac{\pi}{6}$.

The cosine function also has an inverse when its domain is restricted. The arccosine function is the inverse of the cosine function. We write the inverse cosine function as either $\cos^{-1}x$ or arcos x. The restricted domain of $f(x) = \cos x$ is $0 \le x \le \pi$, and the range is $-1 \le \cos x \le 1$. The arccosine function is the inverse of the cosine function on this restricted domain. The domain of the arccosine function is $-1 \le x \le 1$, and the range is $0 \le \arccos x \le \pi$. The restricted domain for the cosine function is different than the restricted domain for the sine function. The equation $y = \arccos x$ *means* $x = \cos y$. The graph of the arccosine function is the graph of the restricted cosine function reflected across the line $y = x$, and is shown in Figure 4.3.

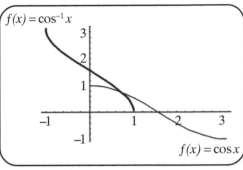

Figure 4.3.

The arccosine function does not inherit the symmetry of the cosine function. The cosine function is symmetric with respect to the y-axis, but we had to restrict its domain so that it could be invertible. In restricting the domain, we removed the symmetry of the function, and as a result, the arccosine function is not symmetric.

Example 2

Find exact values of the following:

a. $\cos^{-1}(-1)$

b. $\cos^{-1}\left(-\frac{1}{2}\right)$

Solution:

a. There are infinitely many angles whose cosine is -1, but we are looking for the unique angle that lies in the restricted domain of the arccosine function: $0 \le \arccos x \le \pi$. $\cos^{-1}(-1) = \pi$.

b. The range of the arccosine function corresponds to angles that lie in either Quadrant I (where the cosine is positive) or Quadrant II (where the cosine is negative).

   To find $\cos^{-1}\left(-\frac{1}{2}\right)$, we need to first find the reference angle whose cosine is $\frac{1}{2}$: $\cos^{-1}\frac{1}{2} = \frac{\pi}{3}$.

   Next, put the angle in the correct quadrant:

   $\cos^{-1}\left(-\frac{1}{2}\right) = \frac{2\pi}{3}$.

From the properties of inverses, we have the following identities:

$$\sin^{-1}(\sin x) = x \text{ if } -\tfrac{\pi}{2} \le x \le \tfrac{\pi}{2} \qquad \sin(\sin^{-1} x) = x \text{ if } -1 \le x \le 1$$

$$\cos^{-1}(\cos x) = x \text{ if } 0 \le x \le \pi \qquad \cos(\cos^{-1} x) = x \text{ if } -1 \le x \le 1$$

We can also mix the arcsine and arccosine functions together, as we will see in the next example.

Example 3

Find exact values for the following:

a. $\sin^{-1}\left(\cos\left(-\tfrac{\pi}{3}\right)\right)$

b. $\cos\left(\sin^{-1}\left(-\tfrac{2}{3}\right)\right)$

Solution: Evaluate each expression starting with the innermost expression. Pay attention to the domain and range of the functions:

a. $\sin^{-1}\left(\cos\left(-\tfrac{\pi}{3}\right)\right) = \sin^{-1}\left(\tfrac{1}{2}\right) = \tfrac{\pi}{6}$

b. This problem involves an angle that we are not very familiar with. In order to evaluate this expression we will need to use a

right triangle. Let $\theta = \sin^{-1}\left(-\tfrac{2}{3}\right)$. This means that $\sin\theta = -\tfrac{2}{3}$, and θ lies in Quadrant IV. We need the value of cos θ. We can construct a right triangle that has a hypotenuse of length 3 and a leg of length 1, as shown in Figure 4.4. We can use the Pythagorean Theorem to determine the length of the other leg:

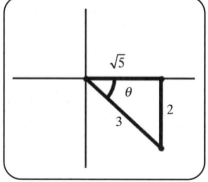

Figure 4.4.

$a = \sqrt{3^2 - 2^2} = \sqrt{9-4} = \sqrt{5}$.

Using this right triangle, we

see that $\cos\theta = \dfrac{\sqrt{5}}{3}$; $\cos\left(\sin^{-1}\left(-\tfrac{2}{3}\right)\right) = \cos\theta = \dfrac{\sqrt{5}}{3}$.

Lesson 4-2 Review

Find exact values for the following:

1. $\sin^{-1}\left(-\frac{\sqrt{3}}{2}\right)$ 　　　 3. $\cos^{-1}\left(\sin\left(-\frac{\pi}{6}\right)\right)$ 　　　 5. $\cos^{-1}\left(\cos\left(\frac{2\pi}{3}\right)\right)$

2. $\cos^{-1}\left(-\frac{\sqrt{3}}{2}\right)$ 　　　 4. $\sin\left(\cos^{-1}\left(-\frac{3}{5}\right)\right)$

Lesson 4-3: The Other Inverse Trigonometric Functions

　　When we restrict the domains of the other four trigonometric functions, they will also be invertible. The arctangent function is the inverse of the tangent function. We write the inverse tangent function as either $\tan^{-1} x$ or arctan x. The restricted domain for the tangent function is $-\frac{\pi}{2} \leq x \leq \frac{\pi}{2}$, and the range of the tangent function over this restricted domain is $(-\infty, \infty)$. The arctangent function is the inverse of the tangent function over this restricted domain. The domain of the arctangent function is $(-\infty, \infty)$, and the range is

$-\frac{\pi}{2} \leq \arctan x \leq \frac{\pi}{2}$. Notice that the restricted domain for the tangent function is the same as the restricted domain for the sine function. The tangent function has vertical asymptotes at $x = \frac{\pi}{2}$ and $x = -\frac{\pi}{2}$. That means that the inverse tangent function will have horizontal asymptotes at $y = \frac{\pi}{2}$ and

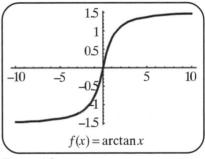

Figure 4.5.

$y = -\frac{\pi}{2}$. The equation $y = \arctan x$
means $x = \tan y$. The graph of the arctangent function is the reflection of the (restricted) tangent function across the line $y = x$, and its graph is shown in Figure 4.5.

　　The inverse secant function is most commonly written as $\sec^{-1} x$. Most calculators do not have (or need) an inverse secant function, because the inverse secant function can be expressed in terms of the arccosine function. First, rewrite the equation $y = \sec^{-1} x$ to remove the reference to the inverse of the function. The equation $y = \sec^{-1} x$ *means* $x = \sec y$. Now we can rewrite this equation in terms of the cosine function and then solve for y again: $y = \sec^{-1} x$

$$x = \sec y$$

$$x = \frac{1}{\cos y}$$

$$\cos y = \frac{1}{x}$$

$$y = \cos^{-1} \frac{1}{x}$$

From this we have the following identity: $\sec^{-1} x = \cos^{-1} \frac{1}{x}$.

The restricted domain for the secant function is $0 \le x \le \pi$ with $x \ne \frac{\pi}{2}$, and the range of the secant function is $(-\infty, -1] \cup [1, \infty)$. The domain of the inverse secant function is $(-\infty, -1] \cup [1, \infty)$, and the range is $0 \le x \le \pi$ with $x \ne \frac{\pi}{2}$. The secant function has a vertical asymptote at $x = \frac{\pi}{2}$, so $y = \frac{\pi}{2}$ will be a horizontal asymptote for the inverse secant function. The graph of the inverse secant function is the reflection of the restricted secant function across the line $y = x$, and its graph is shown in Figure 4.6.

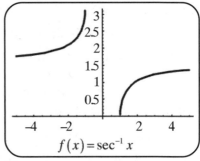

$f(x) = \sec^{-1} x$

Figure 4.6.

The inverse cosecant function is most commonly written as $\csc^{-1} x$. As we discussed with the inverse secant function, calculators do not need an inverse cosecant function, because the inverse cosecant function can be expressed in terms of the arcsine function by following the same approach that we used to rewrite the inverse secant function: $y = \csc^{-1} x$

$$x = \csc y$$

$$x = \frac{1}{\sin y}$$

$$\sin y = \frac{1}{x}$$

$$y = \sin^{-1}\frac{1}{x}$$

From this we have another identity: $\csc^{-1} x = \sin^{-1}\frac{1}{x}$.

The restricted domain for the cosecant function is $-\frac{\pi}{2} \le x \le \frac{\pi}{2}$ with $x \ne 0$, and the range of the cosecant function is $(-\infty, -1] \cup [1, \infty)$. The domain of the inverse cosecant function is $(-\infty, -1] \cup [1, \infty)$, and the range is $-\frac{\pi}{2} \le x \le \frac{\pi}{2}$ with $x \ne 0$. The cosecant function has a vertical asymptote at $x = 0$, so $y = 0$ will be a horizontal asymptote for the inverse cosecant function. The graph of the inverse cosecant function is the reflection of the restricted cosecant function across the line $y = x$, and its graph is shown in Figure 4.7.

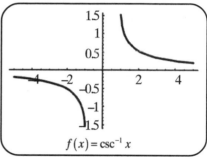

$f(x) = \csc^{-1} x$

Figure 4.7.

We can establish the identity $\cot^{-1} x = \tan^{-1}\frac{1}{x}$ in the same way that we established the identities $\sec^{-1} x = \cos^{-1}\frac{1}{x}$ and $\csc^{-1} x = \sin^{-1}\frac{1}{x}$. The restricted domain for the cotangent function is $0 \le x \le \pi$, and the range of the cotangent function is $(-\infty, \infty)$. The domain of the inverse cotangent function is $(-\infty, \infty)$ and its range satisfies the inequality $0 \le \cot^{-1} x \le \pi$. The cotangent function has vertical asymptotes at $x = 0$ and $x = \pi$. That means that the inverse tangent function will have horizontal asymptotes at $y = 0$ and $y = \pi$. The graph of the inverse cotangent function is the reflection of the restricted cotangent function across the line $y = x$, and its graph is shown in Figure 4.8.

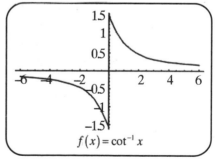

$f(x) = \cot^{-1} x$

Figure 4.8.

Lesson 4-4: Applications

We can evaluate expressions that involve a combination of trigono-metric and inverse trigonometric functions, as we will see in the first example.

Example 1

Find the exact value of the following expressions:

a. $\csc\left[\cos^{-1}\left(-\frac{\sqrt{3}}{2}\right)\right]$

b. $\sec\left[\tan^{-1}(-1)\right]$

c. $\tan\left[\cos^{-1}\left(-\frac{2}{3}\right)\right]$

Solution: If the problem involves the special angles, evaluate the inverse function exactly. Otherwise, use a right triangle and the Pythagorean Theorem to evaluate the expression.

a. $\csc\left[\cos^{-1}\left(-\frac{\sqrt{3}}{2}\right)\right]=\csc\left(\frac{5\pi}{6}\right)=2$

b. $\sec\left[\tan^{-1}(-1)\right]=\sec\left(-\frac{\pi}{4}\right)=\sqrt{2}$

c. $\tan\left[\cos^{-1}\left(-\frac{2}{3}\right)\right]$: For this problem, we are not working with a special angle, so we will need to draw a right tri-angle, as shown in Figure 4.9.

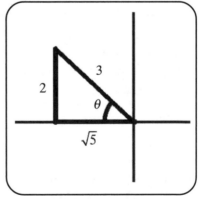

Let $\theta=\cos^{-1}\left(-\frac{2}{3}\right)$.

Then $\tan\theta=-\frac{\sqrt{5}}{2}$, and

$\tan\left[\cos^{-1}\left(-\frac{2}{3}\right)\right]=-\frac{\sqrt{5}}{2}$.

Figure 4.9

The key to evaluating these types of expressions is to know the trigonometric functions for the special angles, or to use a right triangle and the Pythagorean Theorem.

Lesson 4-4 Review

Find the exact value of the following expressions:

1. $\sec\left[\sin^{-1}\left(\frac{2\sqrt{5}}{5}\right)\right]$

2. $\sin\left[\tan^{-1}(-3)\right]$

3. $\cot\left[\cos^{-1}\left(-\frac{\sqrt{3}}{3}\right)\right]$

Answer Key

Lesson 4-1 Review

1. $f(x) = 3x - 4$: $f^{-1}(x) = \frac{1}{3}(x + 4)$

Lesson 4-2 Review

1. $\sin^{-1}\left(-\frac{\sqrt{3}}{2}\right) = -\frac{\pi}{3}$

2. $\cos^{-1}\left(-\frac{\sqrt{3}}{2}\right) = \frac{5\pi}{6}$

3. $\cos^{-1}\left(\sin\left(-\frac{\pi}{6}\right)\right) = \cos^{-1}\left(-\frac{1}{2}\right) = \frac{2\pi}{3}$

4. $\sin\left(\cos^{-1}\left(-\frac{3}{5}\right)\right) = \frac{4}{5}$

5. $\cos^{-1}\left(\cos\left(\frac{2\pi}{3}\right)\right) = \frac{2\pi}{3}$

Lesson 4-4 Review

1. $\sec\left[\sin^{-1}\left(\frac{2\sqrt{5}}{5}\right)\right] = \sqrt{5}$

2. $\sin\left[\tan^{-1}(-3)\right] = -\frac{3}{\sqrt{10}}$

3. $\cot\left[\cos^{-1}\left(-\frac{\sqrt{3}}{3}\right)\right] = -\frac{\sqrt{2}}{2}$

Identities

One reason why mathematics is so important is that it uses systematic methods for solving problems. Another skill that is developed in mathematics is the ability to prove statements or theorems. Proving trigonometric identities is one way to practice this skill.

An *identity* is an equation that is valid under all conditions. In Chapter 2, we established two important trigonometric identities:

$\cos^2 \theta + \sin^2 \theta = 1$ and $1 + \tan^2 \theta = \sec^2 \theta$.

In this chapter, we will establish many more trigonometric identities. We will use these identities to simplify functions and find exact values of the trigonometric functions at angles besides the special angles we discussed in Chapter 1.

Lesson 5-1: Trigonometric Identities

A trigonometric identity is an equation that usually has trigonometric expressions on both sides of the equality. To prove a trigonometric identity, start with one side of the equation and transform it into the other side. Typically, it is best to start with the more complicated side of the equation. You can transform a trigonometric expression by writing the trigonometric functions in terms of sine and cosine functions. Sometimes it helps to multiply an expression by 1, disguised as a ratio of identical trigonometric expressions, such as $\frac{\sec\theta}{\sec\theta}$. You can also use previously established identities to help simplify a trigonometric expression. Keep in mind that the goal is to turn one side of the equation into the other side. You should *not* move any term from one side of the equality to the other side, and you should *not* square both sides of an equation.

The definitions of the trigonometric functions will be instrumental in establishing some identities. The symmetry of the trigonometric functions may also help establish certain identities. I have summarized the definitions and identities that we have already established in the table that follows.

Definitions	Symmetry
$\tan\theta = \dfrac{\sin\theta}{\cos\theta}$	$\sin(-\theta) = -\sin\theta$
$\cot\theta = \dfrac{\cos\theta}{\sin\theta}$	$\cos(-\theta) = \cos\theta$
$\sec\theta = \dfrac{1}{\cos\theta}$	$\tan(-\theta) = -\tan\theta$
$\csc\theta = \dfrac{1}{\sin\theta}$	$\csc(-\theta) = -\csc\theta$
$\cot\theta = \dfrac{1}{\tan\theta}$	$\sec(-\theta) = \sec\theta$
	$\cot(-\theta) = -\cot\theta$

Example 1

Establish the identity $\sec\theta \cdot \sin\theta = \tan\theta$.

Solution: The left side of the equation is the more complicated of the two. Write the expression on the left in terms of sine and cosine functions, and use the definition of the tangent function:

Start with the left side of the equation $\qquad\qquad \sec\theta \cdot \sin\theta$

Use the definition of the secant function $\qquad \dfrac{1}{\cos\theta} \cdot \sin\theta$

Use the definition of the tangent function $\qquad \tan\theta$

We can follow the transformation as a string of equivalent expressions that start as the left side of the identity and end as the right side of the identity:

$$\sec\theta \cdot \sin\theta = \dfrac{1}{\cos\theta} \cdot \sin\theta = \tan\theta$$

Example 2

Establish the identity $\dfrac{1-\sin\theta}{\cos\theta} = \dfrac{\cos\theta}{1+\sin\theta}$.

Solution: Both sides of this identity are equally complex. As a motivation for how to approach this problem, pretend that this identity is valid. Look at what would happen if we cross-multiplied this proportion: $(1 - \sin\theta)(1 + \sin\theta) = \cos^2\theta$.

The product on the left can be expanded:

$(1 - \sin\theta)(1 + \sin\theta) = 1 - \sin^2\theta$.

Remember the first identity that we proved: $\cos^2\theta + \sin^2\theta = 1$. We can rearrange this equation by subtracting $\sin^2\theta$ from both sides: $\cos^2\theta = 1 - \sin^2\theta$.

From our analysis, we see that the identity $\dfrac{1-\sin\theta}{\cos\theta} = \dfrac{\cos\theta}{1+\sin\theta}$ is simply a variation of the identity $\cos^2\theta + \sin^2\theta = 1$. The simplification that we did was a way to understand why the identity is true. It is not a valid proof of the identity, but it does give some insight into how to construct the proof. In order to prove an identity, however, we cannot use the "cross multiply" approach. We must start with one side of the equation and transform it to the other side. I will start with the left side of the equation and multiply it by the number 1 disguised as a ratio of trigonometric expressions:

Start with the left side of the equation $\dfrac{1-\sin\theta}{\cos\theta}$

Multiply the expression by 1 disguised as $\dfrac{1+\sin\theta}{1+\sin\theta}$

$$\dfrac{1-\sin\theta}{\cos\theta} \cdot \dfrac{1+\sin\theta}{1+\sin\theta}$$

Expand the numerator, but keep the denominator factored.

$$\dfrac{1-\sin^2\theta}{\cos\theta(1+\sin\theta)}$$

Rearrange the identity $\cos^2\theta + \sin^2\theta = 1$ to give $\cos^2\theta = 1 - \sin^2\theta$

$$\dfrac{\cos^2\theta}{\cos\theta(1+\sin\theta)}$$

Cancel one of the $\cos\theta$ terms in the numerator with the $\cos\theta$ term in the denominator.

$$\frac{\cos\theta}{(1+\sin\theta)}$$

The last expression is the right side of the identity we were trying to establish. We can follow the transformation as a string of equivalent expressions that start as the left side of the identity and end as the right side of the identity:

$$\frac{1-\sin\theta}{\cos\theta} = \frac{1-\sin\theta}{\cos\theta}\cdot\frac{1+\sin\theta}{1+\sin\theta} = \frac{1-\sin^2\theta}{\cos\theta(1+\sin\theta)} = \frac{\cos^2\theta}{\cos\theta(1+\sin\theta)}$$

$$= \frac{\cos\theta}{(1+\sin\theta)}$$

Example 3

Establish the identity $\tan\theta + \cot\theta = \sec\theta\ \csc\theta$.

Solution: Write the expression on the left in terms of sine and cosine functions, and then simplify:

Start with the left side of the equation $\tan\theta + \cot\theta$

Write the tangent and cotangent functions in terms of sine and

cosine functions
$$\frac{\sin\theta}{\cos\theta} + \frac{\cos\theta}{\sin\theta}$$

Get a common denominator
$$\frac{\sin\theta}{\cos\theta}\cdot\frac{\sin\theta}{\sin\theta} + \frac{\cos\theta}{\sin\theta}\cdot\frac{\cos\theta}{\cos\theta}$$

Add the two fractions together
$$\frac{\sin^2\theta + \cos^2\theta}{\cos\theta\sin\theta}$$

Use the identity $\sin^2\theta + \cos^2\theta = 1$ to simplify the numerator

$$\frac{1}{\cos\theta\sin\theta}$$

Use the definitions of the secant and cosecant functions
$$\sec\theta\ \csc\theta$$

The last expression is the right side of the identity we were trying to establish. We can follow the transformation as a string of equivalent expressions that start as the left side of the identity and end as the right side of the identity:

$$\tan\theta + \cot\theta = \frac{\sin\theta}{\cos\theta} + \frac{\cos\theta}{\sin\theta} = \frac{\sin\theta}{\cos\theta}\cdot\frac{\sin\theta}{\sin\theta} + \frac{\cos\theta}{\sin\theta}\cdot\frac{\cos\theta}{\cos\theta}$$

$$= \frac{\sin^2\theta + \cos^2\theta}{\cos\theta\sin\theta} = \frac{1}{\cos\theta\sin\theta} = \sec\theta\csc\theta$$

Lesson 5-1 Review

Establish the following identities:

1. $\csc\theta - \cot\theta = \dfrac{\sin\theta}{1+\cos\theta}$

2. $\dfrac{\sin\theta\cos\theta}{\cos^2\theta - \sin^2\theta} = \dfrac{\tan\theta}{1-\tan^2\theta}$

Lesson 5-2: Sum and Difference Formulas

There are only a few acute angles for which we can determine the exact values of the trigonometric functions: the special angles that were introduced in Chapter 2. As a reminder, these angles are $\frac{\pi}{6}$, $\frac{\pi}{4}$, $\frac{\pi}{3}$, and $\frac{\pi}{2}$. Using the idea of reference angles, we were able to expand this list, but there are many gaps. For example, we do not know the exact value of $\tan\frac{\pi}{12}$. The sum and difference formulas are identities that involve the sum or difference of two angles. These identities can be used to evaluate trigonometric functions that can be written as the sum or difference of angles whose sine and cosine values are known exactly. In other words, we will be able to find the exact values of the trigonometric functions for angles that can be expressed as a sum or difference of the angles given in the original list presented in Chapter 2.

The sum and difference formulas for the cosine function are:

$$\cos(\alpha + \beta) = \cos\alpha\cos\beta - \sin\alpha\sin\beta$$
$$\cos(\alpha - \beta) = \cos\alpha\cos\beta + \sin\alpha\sin\beta$$

We will see how to use these formulas in the next example.

Example 1

Find the exact value of $\cos\frac{\pi}{12}$.

Solution: We need to write the angle $\frac{\pi}{12}$ as a combination of the special angles presented in Chapter 2. Writing the given angle in terms of the special angles is usually the most difficult step in these problems.

We can write $\frac{\pi}{12}$ in terms of $\frac{\pi}{3}$ and $\frac{\pi}{4}$ as: $\frac{\pi}{12} = \frac{\pi}{3} - \frac{\pi}{4}$. We can use the difference formula for the cosine function to answer the question:

$$\cos\frac{\pi}{12} = \cos\left(\frac{\pi}{3} - \frac{\pi}{4}\right) = \cos\frac{\pi}{3}\cos\frac{\pi}{4} + \sin\frac{\pi}{3}\sin\frac{\pi}{4}$$

$$= \frac{1}{2} \cdot \frac{\sqrt{2}}{2} + \frac{\sqrt{3}}{2} \cdot \frac{\sqrt{2}}{2} = \frac{\sqrt{2} + \sqrt{6}}{4}$$

The sum and difference formulas for the sine function are:

$$\sin(\alpha + \beta) = \sin\alpha\cos\beta + \cos\alpha\sin\beta$$
$$\sin(\alpha - \beta) = \sin\alpha\cos\beta - \cos\alpha\sin\beta.$$

As we saw in the previous example, using these formulas are fairly straightforward. The key is to figure out how to write the given angle in terms of angles for which we know the exact values of the trigonometric functions. We can add $\frac{\pi}{12}$ to this list, as we saw in the previous example. Our list of angles now includes $\frac{\pi}{12}$, $\frac{\pi}{6}$, $\frac{\pi}{4}$, $\frac{\pi}{3}$, and $\frac{\pi}{2}$.

Example 2

Find the exact value of $\sin\frac{\pi}{12}$.

Solution: We need to write the angle $\frac{\pi}{12}$ as a combination of the angles for which we know the exact values of the trigonometric functions. As we saw in the previous example, $\frac{\pi}{12} = \frac{\pi}{3} - \frac{\pi}{4}$. We can use the difference formula for the sine function to answer the question:

$$\sin\tfrac{\pi}{12} = \cos\left(\tfrac{\pi}{3} - \tfrac{\pi}{4}\right) = \sin\tfrac{\pi}{3}\cos\tfrac{\pi}{4} - \cos\tfrac{\pi}{3}\sin\tfrac{\pi}{4}$$

$$= \tfrac{\sqrt{3}}{2} \cdot \tfrac{\sqrt{2}}{2} - \tfrac{1}{2} \cdot \tfrac{\sqrt{2}}{2} = \tfrac{\sqrt{6}-\sqrt{2}}{4}$$

Example 3

Find the exact value of $\sin\tfrac{7\pi}{12}$.

Solution: We need to write the angle $\tfrac{7\pi}{12}$ as a combination of the angles for which we know the exact values of the trigonometric functions: $\tfrac{7\pi}{12} = \tfrac{\pi}{2} + \tfrac{\pi}{12}$. We can use the sum formula for the sine function to answer the question:

$$\sin\tfrac{7\pi}{12} = \cos\left(\tfrac{\pi}{2} + \tfrac{\pi}{12}\right) = \sin\tfrac{\pi}{2}\cos\tfrac{\pi}{12} + \cos\tfrac{\pi}{2}\sin\tfrac{\pi}{12}$$

$$= 1 \cdot \left(\tfrac{\sqrt{2}+\sqrt{6}}{4}\right) + 0 \cdot \tfrac{\sqrt{6}-\sqrt{2}}{4} = \tfrac{\sqrt{2}+\sqrt{6}}{4}$$

Example 3 can also be solved using the relationship $\sin\theta = \cos\left(\tfrac{\pi}{2} - \theta\right)$ and the fact that the cosine function is an even function:

$$\sin\tfrac{7\pi}{12} = \cos\left(\tfrac{\pi}{2} - \tfrac{7\pi}{12}\right) = \cos\left(-\tfrac{\pi}{12}\right) = \cos\tfrac{\pi}{12} = \tfrac{\sqrt{2}+\sqrt{6}}{4}$$

There is often more than one way to solve these problems. Your answers will not be affected by the approach that you take. We made the observation that $\sin\theta = \cos\left(\tfrac{\pi}{2} - \theta\right)$ in Chapter 2. We can actually use the difference formula for the cosine function to establish this relationship:

$$\cos\left(\tfrac{\pi}{2} - \theta\right) = \cos\tfrac{\pi}{2}\cos\theta + \sin\tfrac{\pi}{2}\sin\theta = 0 \cdot \cos\theta + 1 \cdot \sin\theta = \sin\theta$$

Example 4

Suppose that α lies in Quadrant III and $\sin\alpha = -\tfrac{2}{3}$. Also, suppose that β lies in Quadrant II and $\cos\beta = -\tfrac{3}{5}$.

Find $\sin(\alpha + \beta)$ and $\cos(\alpha - \beta)$.

Solution: We will want to use the sum and difference formulas to find $\sin(\alpha + \beta)$ and $\cos(\alpha - \beta)$:

$\sin(\alpha + \beta) = \sin\alpha \cos\beta + \cos\alpha \sin\beta$

$\cos(\alpha - \beta) = \cos\alpha \cos\beta + \sin\alpha \sin\beta$

To find the exact values of these expressions, we will need to know $\sin\alpha$, $\cos\alpha$, $\sin\beta$, and $\cos\beta$. Two of these four values are given to us. We will need to create right triangles and the quadrants of the angles to determine the other two values. Because α lies in

Quadrant III, $\cos\alpha < 0$: $\cos\alpha = -\frac{\sqrt{5}}{3}$. Because β lies in Quadrant

II, $\sin\beta > 0$: $\sin\beta = \frac{4}{5}$. We can now substitute these values into

the sum and difference formulas to answer the question:

$\sin(\alpha + \beta) = \left(-\frac{2}{3}\right)\left(-\frac{3}{5}\right) + \left(-\frac{\sqrt{5}}{3}\right)\left(\frac{4}{5}\right) = \frac{6-4\sqrt{5}}{15}$

$\cos(\alpha - \beta) = \left(-\frac{\sqrt{5}}{3}\right)\left(-\frac{3}{5}\right) + \left(-\frac{2}{3}\right)\left(\frac{4}{5}\right) = \frac{3\sqrt{5}-8}{15}$

If we know the exact value for either the sine or cosine of an angle, we can use a right triangle to find the exact values for *any* of the other trigonometric functions at that angle. Actually, if we know the exact value of one of the trigonometric functions, we can use a right triangle and the Pythagorean Theorem to find the exact values for *all* of the other trigonometric functions. As we saw in the previous example, we only needed to be given the exact value of either the sine or cosine function at each of the two angles in order to solve the problem. With these types of problems, a minimal amount of information is given. The key is to make use of right triangles and the Pythagorean Theorem to find the missing pieces. We can then use that information to find the exact value of various combinations of those two angles using the sum and difference formulas.

There are many identities that involve sum and difference formulas. Follow the same approach to establish these identities as was discussed in the last lesson. Use the sum or difference formulas to expand any terms that involve sums or differences of angles, and then simplify.

Example 5

Establish the identity $\dfrac{\sin(\alpha + \beta)}{\cos\alpha \cos\beta} = \tan\alpha + \tan\beta$.

Solution: Start with the left side of the identity and expand the numerator. Cancel as appropriate and simplify:

Start with the left side of the equation
$$\frac{\sin(\alpha + \beta)}{\cos\alpha \cos\beta}$$

Use the sum formula for the sine function
$$\frac{\sin\alpha \cos\beta + \cos\alpha \sin\beta}{\cos\alpha \cos\beta}$$

Split the fraction into two parts
$$\frac{\sin\alpha \cos\beta}{\cos\alpha \cos\beta} + \frac{\cos\alpha \sin\beta}{\cos\alpha \cos\beta}$$

Cancel the common cosine terms in each fraction

$$\frac{\sin\alpha}{\cos\alpha} + \frac{\sin\beta}{\cos\beta}$$

Use the definition of the tangent function $\quad \tan\alpha + \tan\beta$

The last expression is the right side of the identity we were trying to establish. We can follow the transformation as a string of equivalent expressions that start as the left side of the identity and end as the right side of the identity:

$$\frac{\sin(\alpha + \beta)}{\cos\alpha \cos\beta} = \frac{\sin\alpha \cos\beta + \cos\alpha \sin\beta}{\cos\alpha \cos\beta} = \frac{\sin\alpha \cos\beta}{\cos\alpha \cos\beta} + \frac{\cos\alpha \sin\beta}{\cos\alpha \cos\beta}$$

$$= \frac{\sin\alpha}{\cos\alpha} + \frac{\sin\beta}{\cos\beta} = \tan\alpha + \tan\beta$$

There are also sum and difference formulas for the tangent function. These formulas can be derived using the sum and difference formulas for the sine and cosine functions:

$$\tan(\alpha + \beta) = \frac{\tan\alpha + \tan\beta}{1 - \tan\alpha \tan\beta}$$

$$\tan(\alpha - \beta) = \frac{\tan\alpha - \tan\beta}{1 + \tan\alpha \tan\beta}$$

The sum formula for the tangent function will enable us to prove that the period of the tangent function is π.

$$\tan(\alpha + \pi) = \frac{\tan \alpha + \tan \pi}{1 - \tan \alpha \tan \pi} = \frac{\tan \alpha + 0}{1 - \tan \alpha \cdot 0} = \frac{\tan \alpha}{1} = \tan \alpha$$

Example 6

Suppose that α lies in Quadrant III and $\sin \alpha = -\frac{3}{5}$. Also, suppose that β lies in Quadrant II and $\cos \beta = -\frac{1}{2}$. Find $\tan(\alpha + \beta)$.

Solution: We will use the formula $\tan(\alpha + \beta) = \frac{\tan \alpha + \tan \beta}{1 - \tan \alpha \tan \beta}$. We

need to calculate $\tan \alpha$ and $\tan \beta$. Create two right triangles and use the Pythagorean Theorem to determine the magnitude of $\tan \alpha$ and $\tan \beta$: $|\tan \alpha| = \frac{3}{4}$ and $|\tan \beta| = \sqrt{3}$. Then use the quadrant of the angle to determine the signs of $\tan \alpha$ and $\tan \beta$. Because α lies in Quadrant III, $\tan \alpha > 0$: $\tan \alpha = \frac{3}{4}$. Because β lies in Quadrant II, $\tan \beta < 0$: $\tan \beta = -\sqrt{3}$. We can now substitute these values into the formula for the tangent function to answer the question:

$$\tan(\alpha + \beta) = \frac{\frac{3}{4} + (-\sqrt{3})}{1 - (\frac{3}{4})(-\sqrt{3})} = \frac{3 - 4\sqrt{3}}{4 + 3\sqrt{3}}$$

We can also complicate matters by bringing the inverse trigonometric functions into the mix. When solving these types of problems, remember that $y = f^{-1}(x)$ means that $x = f(y)$.

Example 7

Find the exact value of $\sin\left(\cos^{-1} \frac{2}{3} + \tan^{-1} \frac{3}{5}\right)$.

Solution: We need to use the sine of the sum of two angles. Let $\alpha = \cos^{-1} \frac{2}{3}$ and $\beta = \tan^{-1} \frac{3}{5}$. Then we have:

$$\sin\left(\cos^{-1} \frac{2}{3} + \tan^{-1} \frac{3}{5}\right) = \sin(\alpha + \beta) = \sin \alpha \cos \beta + \cos \alpha \sin \beta$$

We can rewrite the equations for α and β: the equation $\alpha = \cos^{-1} \frac{2}{3}$ means $\cos \alpha = \frac{2}{3}$, and $\beta = \tan^{-1} \frac{3}{5}$ means that $\tan \beta = \frac{3}{5}$.

We can use the Pythagorean Theorem to find $\sin \alpha$, $\cos \alpha$, $\sin \beta$, and $\cos \beta$: $\sin \alpha = \frac{\sqrt{5}}{3}$, $\cos \alpha = \frac{2}{3}$, $\sin \beta = \frac{3}{\sqrt{34}}$, and $\cos \beta = \frac{5}{\sqrt{34}}$.

We can now answer the question:

$$\sin\left(\cos^{-1}\tfrac{2}{3} + \tan^{-1}\tfrac{3}{5}\right) = \sin(\alpha + \beta) = \left(\tfrac{\sqrt{5}}{3}\right)\left(\tfrac{5}{\sqrt{34}}\right) + \left(\tfrac{2}{3}\right)\left(\tfrac{3}{\sqrt{34}}\right) = \tfrac{5\sqrt{5}+6}{3\sqrt{34}}$$

When solving these problems, we must keep in mind the various techniques that are available. If you know the value of one of the trigonometric functions at a particular angle, and if you also know the quadrant of the angle, you can use the Pythagorean Theorem to find the values of the other five trigonometric functions at that angle. Use the sum or difference formulas to simplify as necessary.

Lesson 5-2 Review

1. Find exact values of the following:

    a. $\sin\frac{5\pi}{12}$

    b. $\sin\left(\cos^{-1}\frac{5}{13} - \tan^{-1}\frac{4}{3}\right)$

2. Establish the identity $\sec(\alpha - \beta) = \dfrac{\sec\alpha \sec\beta}{1 + \tan\alpha \tan\beta}$

3. If $\cos\theta = \frac{1}{4}$ and θ lies in Quadrant IV, find the exact value of the following:

    a. $\cos\left(\theta + \frac{\pi}{6}\right)$ b. $\tan\left(\theta - \frac{\pi}{4}\right)$ c. $\sin\left(\theta + \frac{3\pi}{4}\right)$

Lesson 5-3: Twice-Angle Formulas

The twice-angle (or double-angle) formulas for the trigonometric functions can be derived from the sum formulas. The first formula we will derive is the twice-angle formula for the sine function:

$$\sin(2\alpha) = \sin(\alpha + \alpha) = \sin\alpha \, \cos\alpha + \cos\alpha \, \sin\alpha = 2\sin\alpha \, \cos\alpha$$

There are three versions of the twice-angle formula for the cosine function. The first version comes directly from the cosine sum formula:

$$\cos(2\alpha) = \cos(\alpha + \alpha) = \cos\alpha \, \cos\alpha - \sin\alpha \, \sin\alpha = \cos^2 \alpha - \sin^2 \alpha$$

We can transform this equation by using the identity $\sin^2\theta + \cos^2\theta = 1$. We can either make the substitution $\sin^2\theta = 1 - \cos^2\theta$, which gives the formula:

$$\cos 2\alpha = 2\cos^2\alpha - 1$$

or we can use the equation $\cos^2\theta = 1 - \sin^2\theta$, which gives the formula:

$$\cos 2\alpha = 1 - 2\sin^2\alpha$$

The twice-angle formula for the tangent function can be derived from the sum formula for the tangent function:

$$\tan 2\alpha = \tan(\alpha + \alpha) = \frac{\tan\alpha + \tan\alpha}{1 - \tan\alpha\tan\alpha} = \frac{2\tan\alpha}{1 - \tan^2\alpha}$$

We can use these formulas to find the exact values of the trigonometric functions for even more angles. We can also use these formulas to establish more identities.

Example 1

If $\sin\alpha = \frac{2}{3}$ and θ lies in Quadrant II, find the exact value of $\cos 2\theta$.

Solution: Use the third twice-angle formula: $\cos 2\alpha = 1 - 2\sin^2\alpha$:

$$\cos 2\alpha = 1 - \left(\frac{2}{3}\right)^2 = \frac{5}{9}$$

Example 2

If $\tan\theta = \frac{2}{5}$ and θ lies in Quadrant III, find $\sin 2\theta$.

Solution: We will need to use the twice-angle formula for the sine function:

$\sin 2\theta = 2\sin\theta\cos\theta$.

We will need to find $\sin\theta$ and $\cos\theta$. Because θ lies in Quadrant III, $\sin\theta < 0$ and $\cos\theta < 0$. Using the Pythagorean Theorem, we have that $\sin\theta = -\frac{2}{\sqrt{29}}$ and $\cos\theta = -\frac{5}{\sqrt{29}}$. We can now use the twice-angle formula to find $\sin 2\theta$:

$$\sin 2\theta = 2\sin\theta\cos\theta = 2\left(-\frac{2}{\sqrt{29}}\right)\left(-\frac{5}{\sqrt{29}}\right) = \frac{20}{29}$$

Example 3

Establish the identity $\sec 2\theta = \dfrac{\sec^2 \theta}{2 - \sec^2 \theta}$.

Solution: Even though the right side of this identity is more complex than the left side, we can simplify the left side by first rewriting sec 2θ in terms of the cosine function and then using the twice-angle formula for the cosine function. The twice-angle formula for the cosine function has three versions. In this case, the denominator has a 2 in it, so that helps determine which version to use.

Start with the left side of the equation $\qquad$ sec 2θ

Use the definition of the secant function $\qquad \dfrac{1}{\cos 2\theta}$

Use the twice-angle formula $\cos 2\alpha = 2\cos^2 \alpha - 1$

$$\frac{1}{2\cos^2 \alpha - 1}$$

Our goal is to turn this expression into $\dfrac{\sec^2 \theta}{2 - \sec^2 \theta}$.

Our expression has a 1 in the numerator, so we should multiply the numerator and denominator of the expression by $\sec^2 \theta$

$$\frac{1}{2\cos^2 \alpha - 1} \cdot \frac{\sec^2 \theta}{\sec^2 \theta}$$

Use the fact that $\cos^2 \alpha \sec^2 \theta = 1$ to simplify this expression

$$\frac{\sec^2 \theta}{2\cos^2 \alpha \sec^2 \theta - \sec^2 \theta}$$

$$\frac{\sec^2 \theta}{2 - \sec^2 \theta}$$

The last expression is the right side of the identity we were trying to establish. We can follow the transformation as a string of equivalent expressions that start as the left side of the identity and end as the right side of the identity:

$$\sec 2\theta = \frac{1}{\cos 2\theta} = \frac{1}{2\cos^2 \alpha - 1} = \frac{1}{2\cos^2 \alpha - 1} \cdot \frac{\sec^2 \theta}{\sec^2 \theta}$$

$$= \frac{\sec^2 \theta}{2\cos^2 \alpha \sec^2 \theta - \sec^2 \theta} = \frac{\sec^2 \theta}{2 - \sec^2 \theta}$$

The twice-angle formula for the sine function provides a way to re-write the product of a sine and a cosine function into a single expression involving the sine function: $\sin \theta \cos \theta = \frac{1}{2}\sin 2\theta$. The twice-angle formula for the cosine function provides a way to rewrite higher powers of the sine or cosine function into an expression that involves only sine or co-sine functions raised to the first power. Solving the equation

$$\cos 2\theta = 2\cos^2 \theta - 1 \text{ for } \cos^2 \theta \text{ gives: } \cos^2 \theta = \frac{\cos 2\theta + 1}{2},$$

and solving the equation

$$\cos 2\theta = 1 - 2\sin^2 \theta \text{ for } \sin^2 \theta \text{ gives: } \sin^2 \theta = \frac{\cos 2\theta - 1}{2}.$$

Example 4

Find an expression for $\sin^4 \theta$ that only involves sine or cosine functions raised to the first power.

Solution: First, use the rearranged twice-angle formula

$\sin^2 \theta = \dfrac{\cos 2\theta - 1}{2}$ to rewrite $\sin^4 \theta$:

$$\sin^4 \theta = \left(\sin^2 \theta\right)^2 = \left(\frac{\cos 2\theta - 1}{2}\right)^2 = \frac{\cos^2 2\theta - 2\cos 2\theta + 1}{4}$$

At this stage, use the rearranged twice-angle formula

$\cos^2 2\theta = \dfrac{\cos 4\theta + 1}{2}$ to eliminate the $\cos^2 2\theta$ term in the new

expression:

$$\sin^4 \theta = \frac{\cos^2 2\theta - 2\cos 2\theta + 1}{4} = \frac{\left(\frac{\cos 4\theta + 1}{2}\right) - 2\cos 2\theta + 1}{4}$$

$$= \tfrac{1}{8}\cos 4\theta - \tfrac{1}{2}\cos 2\theta + \tfrac{3}{8}$$

Example 5

Find the exact value of $\tan\left[2\cos^{-1}\frac{3}{5}\right]$.

Solution: Let $\theta = \cos^{-1}\frac{3}{5}$. This means that $\cos\theta = \frac{3}{5}$. Then $\tan\left[2\cos^{-1}\frac{3}{5}\right] = \tan 2\theta$. We can use the twice-angle formula for the tangent function to answer the question:

$$\tan\left[2\cos^{-1}\frac{3}{5}\right] = \tan 2\theta = \frac{2\tan\alpha}{1-\tan^2\alpha}.$$

Because $\cos\theta = \frac{3}{5}$, $\tan\theta = \frac{4}{3}$. From this, we see that:

$$\tan\left[2\cos^{-1}\frac{3}{5}\right] = \frac{2\left(\frac{4}{3}\right)}{1-\left(\frac{4}{3}\right)^2} = \frac{\frac{8}{3}}{-\frac{7}{3}} = -\frac{8}{3}.$$

Lesson 5-3 Review

1. If $\cos\theta = \frac{2}{5}$ and θ lies in Quadrant IV, find $\sin 2\theta$.

2. Establish the identity $\cos 2\theta = \dfrac{\cot\theta - \tan\theta}{\cot\theta + \tan\theta}$.

3. Find the exact value of $\sin\left[2\tan^{-1}\frac{5}{8}\right]$.

Lesson 5-4: Half-Angle Formulas

The twice-angle formula for the cosine function can be used to derive the half-angle sine and cosine formulas. Starting with the equation

$$\cos^2\theta = \frac{\cos 2\theta + 1}{2},$$

let $\theta = \dfrac{\alpha}{2}$: $\cos^2\dfrac{\alpha}{2} = \dfrac{\cos\left[2\left(\frac{\alpha}{2}\right)\right]+1}{2} = \dfrac{\cos\alpha + 1}{2}$

If we take the square root of both sides, we have the first half-angle formula:

$$\cos\frac{\alpha}{2} = \pm\sqrt{\frac{\cos\alpha + 1}{2}}$$

We can derive the half-angle formula for the sine function in a similar manner. Start with the equation:

$$\sin^2\theta = \frac{1-\cos 2\theta}{2}$$

and let $\theta = \frac{\alpha}{2}$: $\sin^2\frac{\alpha}{2} = \frac{1-\cos\left[2\left(\frac{\alpha}{2}\right)\right]}{2} = \frac{1-\cos\alpha}{2}$

If we take the square root of both sides, we have the half-angle formula for the sine function:

$$\sin\frac{\alpha}{2} = \pm\sqrt{\frac{1-\cos\alpha}{2}}$$

We can combine the half-angle formulas for the sine and cosine function to derive the half-angle formula for the tangent function:

$$\tan\frac{\alpha}{2} = \pm\sqrt{\frac{1-\cos\alpha}{1+\cos\alpha}}$$

Each of the half-angle formulas has a $\pm$ symbol involved. The sign of the function is determined by the quadrant that $\frac{\alpha}{2}$ lies in.

Example 1

Find the exact value of $\tan\frac{5\pi}{8}$.

Solution: $\frac{5\pi}{8}$ is the same as $\frac{1}{2}\left(\frac{5\pi}{4}\right)$, so we can use the half-angle formula, with $\alpha = \frac{5\pi}{4}$. Also, $\frac{5\pi}{8}$ lies in Quadrant II, so $\tan\frac{5\pi}{8} < 0$:

$$\tan\frac{\alpha}{2} = -\sqrt{\frac{1-\cos\alpha}{1+\cos\alpha}}$$

$$\tan\frac{5\pi}{8} = \tan\left[\frac{1}{2}\left(\frac{5\pi}{4}\right)\right] = -\sqrt{\frac{1-\cos\frac{5\pi}{4}}{1+\cos\frac{5\pi}{4}}}$$

$$\tan\frac{5\pi}{8} = -\sqrt{\frac{1-\cos\frac{5\pi}{4}}{1+\cos\frac{5\pi}{4}}}$$

$$\tan\frac{5\pi}{8} = -\sqrt{\frac{1-\left(-\frac{\sqrt{2}}{2}\right)}{1+\left(-\frac{\sqrt{2}}{2}\right)}}$$

$$\tan\frac{5\pi}{8} = -\sqrt{\frac{2+\sqrt{2}}{2-\sqrt{2}}}$$

We can use the half-angle formulas to establish identities, as we will see in the next example.

Example 2

Establish the identity $\tan\frac{\theta}{2} = \csc\theta - \cot\theta$.

Solution: The method used to establish this identity is not immediately obvious. It may be useful to start with the right side of the identity and try to end up with the left side:

Start with the right side of the identity $\quad\csc\theta - \cot\theta$

Rewrite the functions in terms of sine and cosine functions

$$\frac{1}{\sin\theta} - \frac{\cos\theta}{\sin\theta}$$

Combine the two fractions $\quad\dfrac{1-\cos\theta}{\sin\theta}$

From the twice-angle formula $\sin^2\frac{\theta}{2} = \dfrac{1-\cos\theta}{2}$, we see that

$2\sin^2\frac{\theta}{2} = 1-\cos\theta$, which we can substitute into our expression

$$\frac{2\sin^2\frac{\theta}{2}}{\sin\theta}$$

We can use the identity $\sin 2\alpha = 2 \sin \alpha \cos \alpha$, and let $\alpha = \frac{\theta}{2}$:

$\sin \theta = 2 \sin \frac{\theta}{2} \cos \frac{\theta}{2}$, which we can substitute into our expression:

$$\frac{2 \sin^2 \frac{\theta}{2}}{2 \sin \frac{\theta}{2} \cos \frac{\theta}{2}}$$

Cancel $2 \sin \frac{\theta}{2}$ from the numerator and the denominator

$$\frac{\sin \frac{\theta}{2}}{\cos \frac{\theta}{2}}$$

Use the definition of the tangent function $\tan \frac{\theta}{2}$

The last expression is the left side of the identity we were trying to establish. We can follow the transformation as a string of equivalent expressions that start as the right side of the identity and end as the left side of the identity:

$$\csc \theta - \cot \theta = \frac{1}{\sin \theta} - \frac{\cos \theta}{\sin \theta} = \frac{1 - \cos \theta}{\sin \theta} = \frac{2 \sin^2 \frac{\theta}{2}}{\sin \theta} = \frac{2 \sin^2 \frac{\theta}{2}}{2 \sin \frac{\theta}{2} \cos \frac{\theta}{2}}$$

$$= \frac{\sin \frac{\theta}{2}}{\cos \frac{\theta}{2}} = \tan \frac{\theta}{2}$$

The identity that we just established, $\tan \frac{\theta}{2} = \csc \theta - \cot \theta$, has its advantages over the half-angle formula for the tangent function that was first presented. Finding $\tan \frac{\theta}{2}$ using the equation $\tan \frac{\theta}{2} = \csc \theta - \cot \theta$ does not involve finding square roots and determining which sign to use. We can rework Example 1 using this newly established identity.

Example 3

Find the exact value of $\tan \frac{5\pi}{8}$ using the identity established in Example 2, and compare this answer to the one obtained in Example 1.

Solution: $\tan\frac{5\pi}{8} = \csc\frac{5\pi}{4} - \cot\frac{5\pi}{4} = -\sqrt{2} - 1 = -\left(\sqrt{2}+1\right)$

In Example 1, we showed that $\tan\frac{5\pi}{8} = -\sqrt{\dfrac{2+\sqrt{2}}{2-\sqrt{2}}}$. Numerically,

these results are the same. Using trigonometry, we have estab-

lished that $\sqrt{\dfrac{2+\sqrt{2}}{2-\sqrt{2}}} = \sqrt{2}+1$. We can verify this algebraically by

squaring both sides of this equation and simplifying:

$$\sqrt{\frac{2+\sqrt{2}}{2-\sqrt{2}}} = \sqrt{2}+1$$

Square both sides of this equation $\quad \dfrac{2+\sqrt{2}}{2-\sqrt{2}} = \left(\sqrt{2}+1\right)^{2}$

Expand $\left(\sqrt{2}+1\right)^{2}$ $\quad\quad\quad\quad\quad \dfrac{2+\sqrt{2}}{2-\sqrt{2}} = 3+2\sqrt{2}$

Multiply both sides of the equation by $2-\sqrt{2}$

$$2+\sqrt{2} = \left(3+2\sqrt{2}\right)\left(2-\sqrt{2}\right)$$

Expand $\left(3+2\sqrt{2}\right)\left(2-\sqrt{2}\right)$ $\quad\quad 2+\sqrt{2} = 2+\sqrt{2}$

Trigonometry is filled with identities. Identities provide various ways to calculate trigonometric expressions, which means that the twice-angle and half-angle formulas that we listed are not unique. After working out the details of Example 2, we see that there are at least three ways to find

the exact value of $\tan\frac{\theta}{2}$:

$$\tan\frac{\theta}{2} = \frac{\sin\frac{\theta}{2}}{\cos\frac{\theta}{2}}$$

$$\tan\frac{\theta}{2} = \csc\theta - \cot\theta$$

$$\tan\frac{\theta}{2} = \pm\sqrt{\frac{1-\cos\theta}{1+\cos\theta}}$$

Every identity can be interpreted as an equivalent way to evaluate a trigonometric expression, and under certain circumstances, some expressions are easier to evaluate than others.

Example 4

Find the exact value of $\sin^2\left[\frac{1}{2}\cos^{-1}\frac{5}{13}\right]$.

Solution: Let $\theta = \cos^{-1}\frac{5}{13}$. This means that $\cos\theta = \frac{5}{13}$, and $\sin^2\left[\frac{1}{2}\cos^{-1}\frac{5}{13}\right] = \sin^2\frac{\theta}{2}$. We can use the half-angle formula for the sine function to evaluate this expression:

$$\sin^2\frac{\theta}{2} = \frac{1-\cos\theta}{2} = \frac{1-\frac{5}{13}}{2} = \frac{4}{13}$$

Lesson 5-4 Review

1. If θ lies in Quadrant III, with $\theta > 0$, and $\cot\theta = \frac{3}{4}$, find the exact value of $\sin\frac{\theta}{2}$.

2. Establish the identity: $\cot^2\frac{\theta}{2} = \frac{\sec\theta+1}{\sec\theta-1}$.

3. Find the exact value of $\cos^2\left[\frac{1}{2}\sin^{-1}\frac{5}{13}\right]$.

Lesson 5-5: Sums and Products Formulas

At this point, trigonometry may seem to be all about identities. In this chapter, you have been introduced to many different formulas for simplifying trigonometric expressions. This is the last lesson that involves formulas and identities for various forms of trigonometric expressions. In the next lesson we will use these various identities and formulas to solve trigonometric equations.

We have discussed ways to simplify trigonometric expressions where the argument of the trigonometric function has been more complicated. In essence, the formulas we have developed have helped us analyze horizontal transformations of trigonometric functions. For example, the graph of $\sin\left(\theta + \frac{\pi}{6}\right)$ is the graph of $\sin\theta$ shifted to the left $\frac{\pi}{6}$ units. We can use the formula for the sine of the sum of two angles to rewrite $\sin\left(\theta + \frac{\pi}{6}\right)$:

$$\sin\left(\theta + \frac{\pi}{6}\right) = \sin\theta\,\cos\frac{\pi}{6} + \cos\theta\,\sin\frac{\pi}{6}$$

We have two ways of looking at the same function. There are times when it is more useful to analyze the compact function $f(\theta) = \sin\left(\theta + \frac{\pi}{6}\right)$, and other times when you may have to deal with the expanded form of the function $f(\theta) = \frac{1}{2}\sin\theta + \frac{\sqrt{3}}{2}\cos\theta$. The ability to recognize when a trigonometric identity can simplify a calculation, and understanding that the effect of adding a quantity to the argument of a trigonometric function results in a horizontal shift of the graph of the function, will enable you analyze more complicated trigonometric equations.

Remember that multiplying the argument of a function by a constant stretches or shrinks the graph of the function horizontally. For example, the graph of $\sin 3\theta$ is the graph of $\sin\theta$ contracted horizontally by a factor of 3. The twice-angle and half-angle formulas provide us with another way to look at $\sin 2\theta$ and $\sin\left(\frac{1}{2}\theta\right)$. You should keep in mind that the graph of $\sin 2\theta$ is the graph of $\sin\theta$ contracted horizontally by a factor of 2, and the graph of $\sin\left(\frac{1}{2}\theta\right)$ is the graph of $\sin\theta$ horizontally stretched by a factor of 2. Again, in addition to our graphical analysis of $\sin 2\theta$ and $\sin\left(\frac{1}{2}\theta\right)$, we now have an algebraic way of working with these functions.

I have taken a long time to explain that the sum and difference formulas for the trigonometric functions can help us analyze horizontal translations of functions, and that the twice-angle and half-angle formulas for the trigonometric functions can help us analyze horizontal contractions and stretches of these functions. Our last series of formulas will help us analyze products and sums of the trigonometric functions themselves, rather than products and sums of their arguments.

The product-to-sum formulas relate the product of sine and cosine functions to sums (or differences) of sine and cosine functions. They can be derived from the sum and difference formulas for the sine and cosine functions. Three product-to-sum formulas will be presented. These formulas will help us evaluate expressions of the form $\sin \alpha \sin \beta$, $\cos \alpha \cos \beta$, and $\sin \alpha \cos \beta$, and are derived using the sum and difference formulas for the sine and cosine functions.

Consider the sum and difference formulas for the cosine function:

$\cos(\alpha + \beta) = \cos \alpha \cos \beta - \sin\alpha \ \sin\beta$

$\cos(\alpha - \beta) = \cos \alpha \cos \beta + \sin\alpha \ \sin\beta$

The symmetry of these two equations will enable us to combine them to solve for either $\sin \alpha \sin \beta$ or $\cos \alpha \cos \beta$. If we add the two equations together, the $\sin \alpha \sin \beta$ terms will add out, and on the right side of the equation we will only have the $\cos \alpha \cos \beta$ terms:

$\cos(\alpha - \beta) + \cos(\alpha + \beta) = 2\cos \alpha \cos \beta$

To solve for $\cos \alpha \ \cos \beta$, divide both sides of this equation by 2:

$$\cos\alpha \cos\beta = \tfrac{1}{2}\left[\cos(\alpha - \beta) + \cos(\alpha + \beta)\right]$$

Going back to the sum and difference formulas for the cosine function, if we subtract one equation from the other, the $\cos \alpha \cos \beta$ terms will add out, and on the right side of the equation we will only have the $\sin \alpha \sin \beta$ terms:

$\cos(\alpha - \beta) - \cos(\alpha + \beta) = 2\sin\alpha \sin \beta$

To solve for $\sin\alpha \sin\beta$, divide both sides of this equation by 2:

$$\sin \alpha \sin \beta = \tfrac{1}{2}\left[\cos(\alpha - \beta) - \cos(\alpha + \beta)\right]$$

We can use the sum and difference formulas for the sine function to derive a formula for $\sin \alpha \cos \beta$:

$\sin (\alpha + \beta) = \sin \alpha \cos \beta + \cos \alpha \sin \beta$

$\sin (\alpha - \beta) = \sin \alpha \ \cos \beta - \cos \alpha \ \sin \beta$

If we add these two equations together, the $\cos \alpha \sin \beta$ terms will add out, and we will only have the $\sin \alpha \cos \beta$ terms on the right:

$\sin (\alpha - \beta) + \sin(\alpha + \beta) = 2\sin \alpha \ \cos \beta$

To solve for $\sin \alpha \cos \beta$, divide both sides of this equation by 2:

$$\sin\alpha \cos\beta = \tfrac{1}{2}\left[\sin(\alpha - \beta) + \sin(\alpha + \beta)\right]$$

The sum-to-product formulas provide a relationship between sums or differences of sine functions (or cosine functions) and products of sine and cosine functions:

$$\sin\alpha + \sin\beta = 2\sin\left(\tfrac{\alpha+\beta}{2}\right)\cos\left(\tfrac{\alpha-\beta}{2}\right)$$

$$\sin\alpha - \sin\beta = 2\sin\left(\tfrac{\alpha-\beta}{2}\right)\cos\left(\tfrac{\alpha+\beta}{2}\right)$$

$$\cos\alpha + \cos\beta = 2\cos\left(\tfrac{\alpha+\beta}{2}\right)\cos\left(\tfrac{\alpha-\beta}{2}\right)$$

$$\cos\alpha - \cos\beta = -2\sin\left(\tfrac{\alpha+\beta}{2}\right)\sin\left(\tfrac{\alpha-\beta}{2}\right)$$

Example 1

Establish the identity: $\dfrac{\sin 4\theta + \sin 2\theta}{\cos 4\theta + \cos 2\theta} = \tan 3\theta$.

Solution: Start with the left side of the identity and use the sum-to-product formulas for the sine and cosine functions:

Start with the left side of the identity

$$\frac{\sin 4\theta + \sin 2\theta}{\cos 4\theta + \cos 2\theta}$$

Use the sum-to-product identities for the sine and cosine functions

$$\frac{2\sin 3\theta \cos\theta}{2\cos 3\theta \cos\theta}$$

Cancel the 2 cos θ terms in the numerator and the denominator

$$\frac{\sin 3\theta}{\cos 3\theta}$$

Use the definition of the tangent function $\tan 3\theta$

The last expression is the right side of the identity we were trying to establish. We can follow the transformation as a string of equivalent expressions that start as the left side of the identity and end as the right side of the identity:

$$\frac{\sin 4\theta + \sin 2\theta}{\cos 4\theta + \cos 2\theta} = \frac{2\sin 3\theta \cos\theta}{2\cos 3\theta \cos\theta} = \frac{\sin 3\theta}{\cos 3\theta} = \tan 3\theta$$

Example 2

If θ lies in Quadrant III with $\pi < \theta < \frac{3\pi}{2}$, and $\cos\theta = -\frac{3}{5}$, find

the exact value of $\cos\frac{\theta}{2} - \cos\frac{3\theta}{2}$.

Solution: Use the sum-to-product formula for the difference between cosine functions:

$$\cos\frac{\theta}{2} - \cos\frac{3\theta}{2} = -2\sin\theta\sin\left(-\frac{\theta}{2}\right) = 2\sin\theta\sin\frac{\theta}{2}$$

Using the Pythagorean Theorem, $\sin\theta = -\frac{4}{5}$. To find $\sin\frac{\theta}{2}$, we will need to use the half-angle formula for the sine function:

$\sin\frac{\theta}{2} = \pm\sqrt{\frac{1-\cos\theta}{2}}$. To determine the sign of $\sin\frac{\theta}{2}$, keep in mind

that θ lies in Quadrant III, and $\pi < \theta < \frac{3\pi}{2}$, which means

$\frac{\pi}{2} < \theta < \frac{3\pi}{4}$, and $\sin\frac{\theta}{2} > 0$. We see that $\sin\frac{\theta}{2} = \sqrt{\frac{1-\left(-\frac{3}{5}\right)}{2}} = \sqrt{\frac{4}{5}} = \frac{2}{\sqrt{5}}$,

and $\cos\frac{\theta}{2} - \cos\frac{3\theta}{2} = 2\sin\theta\sin\frac{\theta}{2} = 2\left(-\frac{4}{5}\right)\left(\frac{2}{\sqrt{5}}\right) = -\frac{16}{5\sqrt{5}}$.

Lesson 5-5 Review

1. If θ lies in Quadrant II with $\frac{\pi}{2} < \theta < \pi$, and $\tan\theta = -\frac{2}{3}$, find the

    exact value of $\sin\frac{\theta}{2} - \sin\frac{3\theta}{2}$.

Lesson 5-6: Solving Trigonometric Equations

Most of this chapter has been devoted to proving and using trigono-metric *identities*. These identities will actually help us to solve trigono-metric *equations*. A **trigonometric equation** is an equation that involves trigonometric functions that are only satisfied by specific values of the independent variable. Of course, there is the possibility that a trigonomet-ric equation has no solutions, but the periodic nature of the trigonometric functions mean that when there is a solution to a trigonometric equation, there are usually infinite solutions to the equation. We will learn how to represent all of the solutions to a trigonometric equation mathematically.

Solving a trigonometric equation often involves familiarity with the special angles that were presented in Chapter 2. Also, knowing the zeros of the sine and cosine function can be very helpful when solving trigonometric equations. Remember that sin $\theta = 0$ when θ is an integer multiple of π: $\theta = n\pi$, where n is any integer. In fact, finding the zeros of the sine function is an example of solving the trigonometric equation sin $\theta = 0$. The zeros of the cosine function are the odd half-integer multiples of π: cos $\theta = 0$ when $\theta = \frac{(2n+1)\pi}{2}$, where n is any integer. There are infinite zeros of the sine and cosine function, and they are written in terms of the variable n, which can take on any integer value.

The nature of the trigonometric functions plays a crucial role in solving trigonometric equations. In particular, the periodicity of the functions and whether or not they are bounded are useful properties to keep in mind. The period of the functions $\sin(Ax + B)$, $\cos(Ax + B)$, and $\tan(Ax + B)$ are $\frac{2\pi}{A}$, $\frac{2\pi}{A}$, and $\frac{\pi}{A}$, respectively. Remember that the sine and cosine functions are bounded, and satisfy the inequalities $-1 \leq \sin \theta \leq 1$ and $-1 \leq \cos \theta \leq 1$. Because of this property, the equation sin $\theta = 2$ is an example of a trigonometric equation that has no solutions. The tangent function is not bounded.

The process of solving a trigonometric equation can be simplified by using trigonometric identities. There are several approaches to solving trigonometric equations. You can solve them algebraically or graphically. Graphing calculators can be particularly helpful in solving trigonometric equations. Familiarity with the values of the trigonometric functions for the special angles, though, is extremely important. The first step in solving a trigonometric equation is to find one solution, which will usually be an acute angle, or a reference angle. From the periodic nature of the trigonometric functions, the other solutions to the equation can be found.

Example 1

Solve the equation $\sin\theta = \frac{1}{2}$.

Solution: Remember that $\sin\frac{\pi}{6} = \frac{1}{2}$, so we have one solution to the equation. Because the period of sin θ is 2π, one set of solutions to the equation is of the form $\frac{\pi}{6} + 2n\pi$. Also, the sine function is positive in Quadrant II, and an angle in Quadrant II whose reference angle is $\frac{\pi}{6}$ will also be a solution to the equation, so $\frac{5\pi}{6}$ will also be a solution to the equation $\sin\theta = \frac{1}{2}$. The periodic nature of the sine function means that the angles

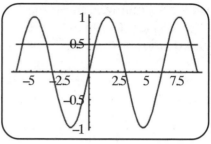

Figure 5.1.

$\frac{5\pi}{6} + 2n\pi$ will also be solutions to the equation $\sin\theta = \frac{1}{2}$. The solutions to the equation $\sin\theta = \frac{1}{2}$ are $\theta = \frac{\pi}{6} + 2n\pi$ or $\theta = \frac{5\pi}{6} + 2n\pi$, where n is any integer. We can also solve this equation graphically. Figure 5.1 shows the graphs of $y = \sin x$ and $y = \frac{1}{2}$. The points where the two functions intersect are the solutions to the equation $\sin\theta = \frac{1}{2}$.

Example 2

Solve the equation $\sin(2\theta + \pi) = \frac{1}{2}$.

Solution: When the argument of the sine function is either $\frac{\pi}{6}$ or $\frac{5\pi}{6}$, the sine function will be $\frac{1}{2}$. Set the argument of the sine function equal to $\frac{\pi}{6}$ or $\frac{5\pi}{6}$ and solve for θ:

$2\theta + \pi = \frac{\pi}{6}$	$2\theta + \pi = \frac{5\pi}{6}$
$2\theta = \frac{\pi}{6} - \pi = -\frac{5\pi}{6}$	$2\theta = \frac{5\pi}{6} - \pi = -\frac{\pi}{6}$
$\theta = -\frac{5\pi}{12}$	$\theta = -\frac{\pi}{12}$

Now, the period of $\sin(2\theta + \pi)$ is π, so the solutions to $\sin(2\theta + \pi) = \frac{1}{2}$ are $\theta = -\frac{5\pi}{12} + n\pi$ or $\theta = -\frac{\pi}{12} + n\pi$, where n is any integer.

Example 3

Find the solutions to the equation $\tan 2\theta + 1 = 0$.

Solution: Isolate the tangent function and then solve for θ

$$\tan 2\theta + 1 = 0$$

Subtract 1 from both sides of the equation $\tan 2\theta = -1$

Solve this equation for 2θ

$$2\theta = \frac{3\pi}{4}$$

$$\theta = \frac{3\pi}{8}$$

Now, $\tan 2\theta$ has period $\frac{\pi}{2}$, so the solutions to $\tan 2\theta + 1 = 0$ are

$\theta = \frac{3\pi}{8} + \frac{n\pi}{2}$, where n is any integer.

Example 4

Find the solutions to the equation $\sin \theta \cos \theta = \frac{1}{2}$.

Solution: In this problem, the product of the sine and cosine functions remind us of the twice angle formula: $\sin 2\theta = 2\sin \theta \cos \theta$. We can multiply both sides of the equation by 2 and use this identity to help solve the equation.

$$\sin \theta \cos \theta = \frac{1}{2}$$

Multiply both sides of the equation by 2 $2 \sin \theta \cos \theta = 1$

Use the twice-angle formula $\sin 2\theta = 1$

Solve this equation for 2θ

$$2\theta = \frac{\pi}{2}$$

Divide both sides of this equation by 2 $\theta = \frac{\pi}{4}$

The period of $\sin 2\theta$ is π, so the solutions to $\sin \theta \cos \theta = \frac{1}{2}$ are

$\theta = \frac{\pi}{4} + n\pi$, where n is any integer.

This problem can also be solved by graphing the functions $y = \sin x \cos x$ and $y = \frac{1}{2}$. The solutions to this problem are the points where the two graphs intersect, as shown in Figure 5.2.

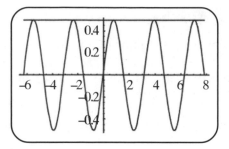

Figure 5.2.

Example 5

Find the solutions to the equation $\cos \theta = \sin \theta$.

Solution: Divide both sides of this equation by $\cos \theta$:

$$\frac{\cos\theta}{\cos\theta} = \frac{\sin\theta}{\cos\theta}$$

We have turned this problem into solving the equation $\tan \theta = 1$. The solutions to this equation are $\theta = \frac{\pi}{4} + n\pi$, where n is any integer.

Example 6

Find the solutions to the equation $\cos 2\theta + 5 \cos \theta + 3 = 0$.

Solution: Use the twice-angle formula, $\cos 2\theta = 2 \cos^2 \theta - 1$, to turn this trigonometric equation into a quadratic equation in $\cos \theta$ which can then be factored:

$2 \cos^2 \theta - 1 + 5 \cos \theta + 3 = 0$

$2 \cos^2 \theta + 5 \cos \theta + 2 = 0$

$(2 \cos \theta + 1)(\cos \theta + 2) = 0$

We can set each factor equal to 0 and solve the original equation:

$(2 \cos \theta + 1) = 0$	$(\cos \theta + 2) = 0$
$\cos\theta = -\frac{1}{2}$	$\cos \theta = -2$
$\theta = \frac{2\pi}{3}$ or $\theta = \frac{4\pi}{3}$	No solutions

The period of $\cos 2\theta + 5 \cos \theta + 3 = 0$ is 2π, so the solutions to $\cos$ $2\theta + 5 \cos \theta + 3 = 0$ are $\theta = \frac{2\pi}{3} + 2n\pi$ or $\theta = \frac{4\pi}{3} + 2n\pi$.

Lesson 5-6 Review

Find the solutions to the following equations:

1. $\cos\left(\theta + \frac{\pi}{3}\right) = \frac{1}{2}$

2. $\sec \frac{3\theta}{2} = -2$

3. $\cos 3\theta + 1 = 0$

4. $\cos 2\theta \sin 2\theta = 0$

5. $\sqrt{\dfrac{1 - \cos\theta}{2}} = 1$

Answer Key
Lesson 5-1 Review

1.
$$\csc\theta - \cot\theta = \frac{1}{\sin\theta} - \frac{\cos\theta}{\sin\theta} = \frac{1 - \cos\theta}{\sin\theta} = \frac{1 - \cos\theta}{\sin\theta} \cdot \frac{1 + \cos\theta}{1 + \cos\theta}$$
$$= \frac{1 - \cos^2\theta}{\sin\theta(1 + \cos\theta)} = \frac{\sin^2\theta}{\sin\theta(1 + \cos\theta)} = \frac{\sin\theta}{1 + \cos\theta}$$

2.
$$\frac{\sin\theta\cos\theta}{\cos^2\theta - \sin^2\theta} = \frac{\sin\theta\cos\theta}{\cos^2\theta - \sin^2\theta} \cdot \frac{\frac{1}{\cos^2\theta}}{\frac{1}{\cos^2\theta}} = \frac{\frac{\sin\theta}{\cos\theta}}{1 - \frac{\sin^2\theta}{\cos^2\theta}} = \frac{\tan\theta}{1 - \tan^2\theta}$$

Lesson 5-2 Review
1. Find exact values of the following:

    a. $\sin\frac{5\pi}{12} = \sin\left(\frac{\pi}{2} - \frac{\pi}{12}\right) = \sin\frac{\pi}{2}\cos\frac{\pi}{12} - \cos\frac{\pi}{2}\sin\frac{\pi}{12} = 1 \cdot \frac{\sqrt{2}+\sqrt{6}}{4} - 0 \cdot \frac{\sqrt{6}-\sqrt{2}}{4}$

        $= \frac{\sqrt{2}+\sqrt{6}}{4}$

b. $\sin\left(\cos^{-1}\frac{5}{13} - \tan^{-1}\frac{4}{3}\right)$: Let $\alpha = \cos^{-1}\frac{5}{13}$ and $\beta = \tan^{-1}\frac{4}{3}$. Then

$\cos\alpha = \frac{5}{13}$, $\sin\alpha = \frac{12}{13}$, $\tan\beta = \frac{4}{3}$, $\sin\beta = \frac{4}{5}$ and $\cos\beta = \frac{3}{5}$. Then:

$$\sin\left(\cos^{-1}\frac{5}{13} - \tan^{-1}\frac{4}{3}\right) = \sin(\alpha - \beta) = \sin\alpha\cos\beta - \cos\alpha\sin\beta$$

$$= \frac{12}{13}\cdot\frac{3}{5} - \frac{5}{13}\cdot\frac{4}{5} = \frac{16}{65}$$

2. $\sec(\alpha - \beta) = \dfrac{1}{\cos(\alpha-\beta)} = \dfrac{1}{\cos\alpha\cos\beta + \sin\alpha\sin\beta} = \dfrac{1}{\cos\alpha\cos\beta + \sin\alpha\sin\beta}\cdot\dfrac{\frac{1}{\cos\alpha\cos\beta}}{\frac{1}{\cos\alpha\cos\beta}}$

$$= \dfrac{\sec\alpha\sec\beta}{1 + \frac{\sin\alpha\sin\beta}{\cos\alpha\cos\beta}} = \dfrac{\sec\alpha\sec\beta}{1 + \tan\alpha\tan\beta}$$

3. a. $\cos\left(\theta + \frac{\pi}{6}\right) = \cos\theta\cos\frac{\pi}{6} - \sin\theta\sin\frac{\pi}{6} = \frac{1}{4}\cdot\frac{\sqrt{3}}{2} - \left(-\frac{\sqrt{15}}{4}\right)\cdot\frac{1}{2} = \frac{\sqrt{3}+\sqrt{15}}{8}$

b. $\tan\left(\theta - \frac{\pi}{4}\right) = \dfrac{\tan\theta - \tan\frac{\pi}{4}}{1 + \tan\theta\tan\frac{\pi}{4}} = \dfrac{-\sqrt{15}-1}{1 - \sqrt{15}}$

c. $\sin\left(\theta + \frac{3\pi}{4}\right) = \sin\theta\cos\frac{3\pi}{4} + \cos\theta\sin\frac{3\pi}{4} = \left(-\frac{\sqrt{15}}{4}\right)\cdot\left(-\frac{\sqrt{2}}{2}\right) + \frac{1}{4}\left(\frac{\sqrt{2}}{2}\right) = \frac{\sqrt{30}+\sqrt{2}}{8}$

Lesson 5-3 Review

1. $\sin 2\theta = 2\sin\theta\cos\theta = 2\left(-\frac{\sqrt{21}}{5}\right)\left(\frac{2}{5}\right) = -\frac{4\sqrt{21}}{25}$

2. $\dfrac{\cot\theta - \tan\theta}{\cot\theta + \tan\theta} = \dfrac{\frac{\cos\theta}{\sin\theta} - \frac{\sin\theta}{\cos\theta}}{\frac{\cos\theta}{\sin\theta} + \frac{\sin\theta}{\cos\theta}} = \dfrac{\frac{\cos\theta}{\sin\theta} - \frac{\sin\theta}{\cos\theta}}{\frac{\cos\theta}{\sin\theta} + \frac{\sin\theta}{\cos\theta}}\cdot\dfrac{\sin\theta\cos\theta}{\sin\theta\cos\theta} = \dfrac{\cos^2\theta - \sin^2\theta}{\cos^2\theta + \sin^2\theta} = \cos 2\theta$

3. Let $\alpha = \tan^{-1}\frac{5}{8}$. Then $\tan\alpha = \frac{5}{8}$, $\cos\alpha = \frac{8}{\sqrt{89}}$, $\sin\alpha = \frac{5}{\sqrt{89}}$,

and $\sin\left[2\tan^{-1}\frac{5}{8}\right] = \sin 2\alpha = 2\sin\alpha\cos\alpha = 2\left(\frac{5}{\sqrt{89}}\right)\left(\frac{8}{\sqrt{89}}\right) = \frac{80}{89}$.

Lesson 5-4 Review

1. Use the formula $\tan\frac{\theta}{2} = \csc\theta - \cot\theta$ to determine the quadrant that $\frac{\theta}{2}$ lies in:

$\tan\frac{\theta}{2} = -\frac{5}{4} - \frac{3}{4} = -2 < 0$, which means that $\frac{\theta}{2}$ is in Quadrant II.

$\sin\frac{\theta}{2} = \sqrt{\frac{1-\cos\theta}{2}} = \sqrt{\frac{1 - \frac{3}{5}}{2}} = \sqrt{\frac{\frac{2}{5}}{2}} = \frac{1}{\sqrt{5}}$

2. $\cot^2 \frac{\theta}{2} = \frac{\cos^2 \frac{\theta}{2}}{\sin^2 \frac{\theta}{2}} = \frac{\frac{1+\cos\theta}{2}}{\frac{1-\cos\theta}{2}} = \frac{1+\cos\theta}{1-\cos\theta} \cdot \frac{\sec\theta}{\sec\theta} = \frac{\sec\theta+1}{\sec\theta-1}$

3. Let $\alpha = \sin^{-1}\frac{5}{13}$. Then $\sin\alpha = \frac{5}{13}$, $\cos\alpha = \frac{12}{13}$, and

$\cos^2\left[\frac{1}{2}\sin^{-1}\frac{5}{13}\right] = \cos^2\frac{\alpha}{2} = \frac{1+\cos\alpha}{2} = \frac{1+\frac{12}{13}}{2} = \frac{25}{26}$.

Lesson 5-5 Review

1. $\sin\frac{\theta}{2} - \sin\frac{3\theta}{2} = 2\sin\left(\frac{\frac{\theta}{2}-\frac{3\theta}{2}}{2}\right)\cos\left(\frac{\frac{\theta}{2}+\frac{3\theta}{2}}{2}\right) = 2\sin\left(-\frac{\theta}{2}\right)\cos\theta = -2\sin\frac{\theta}{2}\cos\theta$

Because $\tan\theta = -\frac{2}{3}$, $\cos\theta = -\frac{3}{\sqrt{13}}$, and

$\sin\frac{\theta}{2} = \sqrt{\frac{1-\cos\theta}{2}} = \sqrt{\frac{1-\left(-\frac{3}{\sqrt{13}}\right)}{2}} = \sqrt{\frac{\sqrt{13}+3}{2\sqrt{13}}}$.

Because $\frac{\pi}{2} \le \theta \le \pi$, we know that $\sin\frac{\theta}{2} > 0$. Substituting in for $\sin\frac{\theta}{2}$, we have:

$\sin\frac{\theta}{2} - \sin\frac{3\theta}{2} = -2\sin\frac{\theta}{2}\cos\theta = -2\left(\sqrt{\frac{\sqrt{13}+3}{2\sqrt{13}}}\right)\left(-\frac{3}{\sqrt{13}}\right) = 6\sqrt{\frac{\sqrt{13}+3}{26\sqrt{13}}} = \frac{6}{13}\sqrt{\frac{13+3\sqrt{13}}{2}}$

Lesson 5-6 Review

1. $\cos\left(\theta+\frac{\pi}{3}\right) = \frac{1}{2}$: $\left(\theta+\frac{\pi}{3}\right) = \frac{\pi}{3}+2\pi n \rightarrow \theta = 0+2\pi n \rightarrow \theta = 2\pi n$

   or $\left(\theta+\frac{\pi}{3}\right) = -\frac{\pi}{3}+2\pi n \rightarrow \theta = -\frac{2\pi}{3}+2\pi n$

2. $\sec\frac{3\theta}{2} = -2$: $\frac{3\theta}{2} = \frac{2\pi}{3}+2\pi n \rightarrow \theta = \frac{4\pi}{9}+\frac{4\pi n}{3}$

   or $\frac{3\theta}{2} = \frac{4\pi}{3}+2\pi n \rightarrow \theta = \frac{8\pi}{9}+\frac{4\pi n}{3}$

3. $\cos 3\theta = -1$: $3\theta = \pi+2\pi n \rightarrow \theta = \frac{\pi}{3}+\frac{2\pi n}{3}$

4. $\frac{1}{2}\sin 4\theta = 0$,

   or $\sin 4\theta = 0$: $4\theta = 0+2\pi n \rightarrow 4\theta = 2\pi n \rightarrow \theta = \frac{2\pi n}{4} = \frac{\pi n}{2}$

5. $\sqrt{\dfrac{1-\cos\theta}{2}} = 1$: Use the identity $\sqrt{\dfrac{1-\cos\theta}{2}} = \pm\sin\dfrac{\theta}{2}$ to substitute into the equation.

$\sin\dfrac{\theta}{2} = \pm 1$

Solve for θ: $\dfrac{\theta}{2} = \dfrac{\pi}{2} + 2\pi n \rightarrow \theta = \pi + 4\pi n$

or $\dfrac{\theta}{2} = -\dfrac{\pi}{2} + 2\pi n \rightarrow \theta = -\pi + 4\pi n$.

Combining these, we have $\theta = \pi + 2n\pi$. You must be careful when using the half-angle formulas: you cannot ignore the $\pm$ sign. You could also square both sides of the equation and solve for θ.

$$\left(\sqrt{\dfrac{1-\cos\theta}{2}}\right)^{2} = 1^{2} \rightarrow \dfrac{1-\cos\theta}{2} = 1 \rightarrow \cos\theta = -1 \rightarrow \theta = \pi + 2n\pi \ .$$

The second method requires fewer steps to give the same result.

Applications of Trigonometric Functions

There is more to trigonometry than proving identities and finding the exact values of trigonometric functions at specific angles. Trigonometry can be used in a variety of situations, such as modeling the number of daylight hours, or calculating the area of a triangular region. In this chapter, we will use trigonometry to solve problems that involve triangles.

Lesson 6-1: Right Triangles

Geometry and trigonometry are closely related. The trigonometric ratios were defined in terms of a right triangle, and the Pythagorean Theorem has been used to solve many of the problems we have seen in this book. It is not surprising that our first applications of trigonometry will focus on right triangles.

Example 1

A trail leads from the parking lot, which has an elevation of 10,000 feet, to a scenic overlook that has an elevation of 14,000 feet. If the length of the trail is 15,000 feet, find the angle of inclination of the trail.

Solution: We can construct a right triangle to fit this situation. Let θ represent the angle of inclination. The length of the hypotenuse is 15,000, and the length of the side opposite the angle of inclination, θ, is 4,000 feet. Using this information, we see that

$\sin\theta = \frac{4,000}{14,000} = \frac{2}{7}$, so $\theta = \sin^{-1}\frac{2}{7}$. Using a calculator, $\theta = 16.60°$.

Example 2

A plane takes off due north and travels for 20 minutes at a speed of 450 mph. It then heads west for 50 minutes at a speed of 540 mph. How far away from the airport is the plane 70 minutes after it takes off?

Solution: Because the directions west and north are perpendicular to each other, the plane is traveling along two legs of a right triangle. The distance from the airport will be the length of the hypotenuse of this right triangle. To find the length of the hypotenuse, we will need to calculate the distance traveled in each direction and use the Pythagorean Theorem. The speed of the plane is given in miles per hour, and the times are given in minutes. We will need to convert the minutes to hours and use the fact that rate times time equals distance, or $r \cdot t = d$, to find the distance:

Distance north: $d = 450 \cdot \frac{20}{60} = 150$ miles

Distance west: $d = 540 \cdot \frac{50}{60} = 450$ miles

The length of the hypotenuse is: $c = \sqrt{a^2 + b^2}$

$$c = \sqrt{150^2 + 450^2} = 150\sqrt{10} \approx 474$$

The plane is approximately 474 miles from the airport after 70 minutes.

Example 3

A rectangle is inscribed in a semicircle of radius 3, as shown in Figure 6.1. Find an equation to represent the area of the rectangle as a function of the angle θ.

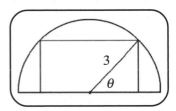

Figure 6.1.

Solution: We can use the trigonometric ratios to find the lengths of the sides of the rectangles in terms of the angle θ. The height of the rectangle is $3 \sin \theta$, and the base of the rectangle is twice the length of the other leg of the right triangle. The length of the other leg of the right triangle is $3 \cos \theta$. The area of the rectangle is $A_{\text{rectangle}} = (3 \sin \theta)(2 \cdot 3 \cos \theta) = 18 \sin \theta \cos \theta$.

We can use the identity $2 \sin \theta \cos \theta = \sin 2\theta$ to simplify our equation for the area of the rectangle: $A_{rectangle} = 9 \sin 2\theta$.

Lesson 6-1 Review

1. A 40 ft. ladder leans against a building. If the bottom of the ladder is 10 feet from the base of the building, what is the angle formed by the ladder and the building?

2. A tree casts a shadow 500 feet long. Find the height of the tree if the angle of elevation of the sun is 23.6°.

3. From a point on the ground 300 ft. from the base of a building, an observer finds that the angle of elevation to the top of the building is 20°, and that the angle of elevation to the top of an antenna on top of the building is 24°. Find the height of the building and the length of the antenna.

Lesson 6-2: The Law of Sines

The trigonometric ratios are based on the relationships between the lengths of the sides of a right triangle. Not all triangles are right triangles, so we will need to develop a strategy for working with all kinds of triangles. The *Law of Sines* will help us relate the sides and angles of any type of triangle.

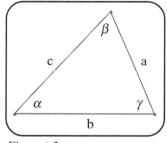

Figure 6.2.

A triangle will either have one right angle and two acute angles, three acute angles, or one obtuse angle and two acute angles. Given a triangle with angles α, β, and γ, let a, b, and c represent the lengths of the sides opposite angles α, β, and γ as shown in Figure 6.2. The **Law of Sines** gives the relationship between the angles and the lengths of the sides:

$$\frac{\sin \alpha}{a} = \frac{\sin \beta}{b} = \frac{\sin \gamma}{c}$$

We can also make use of the fact that the interior angles of a triangle add up to 180°, giving us the equation $\alpha + \beta + \gamma = 180°$. The Law of Sines can be used to solve problems in which either one side and two angles, or two sides and the angle opposite one of them are known. In problems where the lengths of two sides and the measure of one opposite angle are known, it is possible that there are no solutions, one solution, or two solutions. In this situation, we will need to explore all possibilities.

Example 1

Find the length of the other two sides of the triangle shown in Figure 6.3.

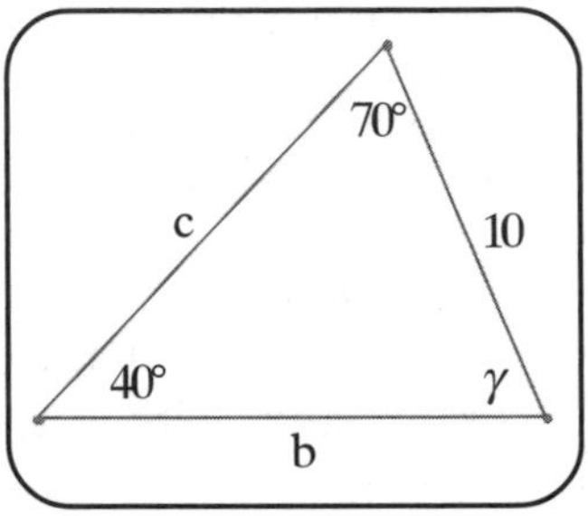

Solution: Use the Law of Sines to first find b:

$$\frac{\sin \alpha}{a} = \frac{\sin \beta}{b}$$

$$\frac{\sin 40°}{10} = \frac{\sin 70°}{b}$$

$$b = \frac{10 \sin 70°}{\sin 40°} \approx 14.62$$

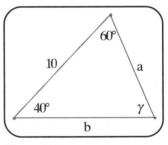

Figure 6.3.

In order to find c, we must know γ:

$\gamma = 180° - \alpha - \beta$

$\gamma = 180° - 40° - 70° = 70°$

Because $\gamma = \beta$, we have an isosceles triangle, and $c = b \approx 14.62$.

Example 2

Find the length of the other two sides of the triangle shown in Figure 6.4.

Solution: Once we know the measure of γ we will be able to use the Law of Sines:

$\gamma = 180° - \alpha - \beta$

$\gamma = 180° - 40° - 60° = 80°$

Now we can find the lengths of the other sides using the Law of Sines. First, find a:

Figure 6.4.

$$\frac{\sin \alpha}{a} = \frac{\sin \gamma}{c}$$

$$\frac{\sin 40°}{a} = \frac{\sin 80°}{10}$$

$$a = \frac{10\sin 40°}{\sin 80°} \approx 6.53$$

Next, find b:

$$\frac{\sin \beta}{b} = \frac{\sin \gamma}{c}$$

$$\frac{\sin 60°}{b} = \frac{\sin 80°}{10}$$

$$b = \frac{10\sin 60°}{\sin 80°} \approx 8.79$$

When the Law of Sines is used to solve for the measure of an angle in a triangle, it is possible that there are two solutions: one angle in Quadrant I and the other angle in Quadrant II. In this case, both solutions must be examined further.

Example 3

Find the measures of the other two angles and the length of the other side of the triangle shown in Figure 6.5.

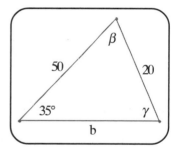

Solution: Using the Law of Sines, we have:

$$\frac{\sin \alpha}{a} = \frac{\sin \gamma}{c}$$

$$\frac{\sin 35°}{20} = \frac{\sin \gamma}{50}$$

Figure 6.5.

$$\sin \gamma = \frac{50\sin 35°}{20} \approx 1.43$$

Now, the sine function must satisfy the inequality $-1 \le \sin \gamma \le 1$, so there is no angle whose sine is 1.43. Therefore, there is no solution to this problem, and such a triangle cannot exist.

Example 4

Find the measures of the other two angles and the length of the other side of the triangle shown in Figure 6.6.

Solution: Using the Law of Sines, we have:

$$\frac{\sin\alpha}{a} = \frac{\sin\gamma}{c}$$

$$\frac{\sin 45°}{80} = \frac{\sin\gamma}{100}$$

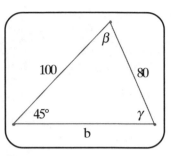

$$\sin\gamma = \frac{100\sin 45°}{80} \approx 0.884$$

Figure 6.6.

$$\gamma = \sin^{-1}(0.884) \approx 62.1°$$

Now, if $\gamma = 62.1°$, then from the relationship $\alpha + \beta + \gamma = 180°$, we see that $\beta = 180° - \gamma - \alpha$, so $\beta = 180° - 62.1° - 45° = 72.9°$. Using the Law of Sines, we can now find b:

$$\frac{\sin\alpha}{a} = \frac{\sin\beta}{b}$$

$$\frac{\sin 45°}{80} = \frac{\sin 72.9°}{b}$$

$$b = \frac{80\sin 72.9°}{\sin 45°} \approx 108.1$$

We can check to make sure that our numbers are consistent. Notice that $\alpha < \gamma < \beta$, and $a < c < b$, so the lengths of the sides are consistent with the relative sizes of the angles. It is a good idea to check your answers to make sure that they make sense. We have found one such triangle that satisfies the given information, but there may be other solutions to this problem. It is important to recognize that there is an angle in Quadrant II whose sine is 0.884: $\gamma \approx 117.9°$. Based on that possibility, we need to see if we can create another triangle that satisfies the given information.

If $\gamma \approx 117.9°$, then from the relationship $\beta = 180° - \gamma - \alpha$, we have $\beta = 180° - 117.9° - 45° = 17.1°$. Using the Law of Sines, we can now find b:

$$\frac{\sin\alpha}{a} = \frac{\sin\beta}{b}$$

$$\frac{\sin 45°}{80} = \frac{\sin 17.1°}{b}$$

$$b = \frac{80\sin 17.1°}{\sin 45°} \approx 33.3$$

Again, we should check to make sure that our numbers make sense. Notice that with this solution, $\beta < \alpha < \gamma$, and $b < a < c$, so the lengths of the sides are consistent with the relative sizes of the angles. In this problem, there are two triangles that satisfy the given conditions.

Example 5

Find the measures of the other two angles and the length of the other side of the triangle shown in Figure 6.7.

Solution: Using the Law of Sines, we have:

$$\frac{\sin\alpha}{a} = \frac{\sin\gamma}{c}$$

$$\frac{\sin 45°}{14} = \frac{\sin\gamma}{10}$$

$$\sin\gamma = \frac{10\sin 45°}{14} \approx 0.505$$

$$\gamma = \sin^{-1}(0.505) \approx 27.7°$$

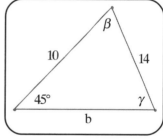

Figure 6.7.

Now, if $\gamma = 27.7°$, then from the relationship $\alpha + \beta + \gamma = 180°$, we see that $\beta = 180° - \gamma - \alpha$, so $\beta = 180° - 27.7° - 45° = 107.3°$. Using the Law of Sines, we can now find b:

$$\frac{\sin\alpha}{a} = \frac{\sin\beta}{b}$$

$$\frac{\sin 45°}{14} = \frac{\sin 107.3°}{b}$$

$$b = \frac{14\sin 107.3°}{\sin 45°} \approx 18.9$$

We can check to make sure that our numbers are consistent. Notice that $\gamma < \alpha < \beta$, and $c < a < b$, so the lengths of the sides are consistent with the relative sizes of the angles. We have found one such triangle that satisfies the given information, but there may be another solution to this problem: $\gamma \approx 152.3°$ is an angle in

Quadrant II whose sine is 0.505. We may be able to use this other angle to create a second triangle that satisfies the given information. Now, if $\gamma \approx 152.3°$, and $\alpha = 45°$ then the sum of these two angles is greater than 180°, which is the sum of the interior angles of a triangle. We cannot use this second angle to create a triangle, and this problem only has one solution.

Surveyors use a method called *triangulation* to measure the distance between two points. The distance between two surveying stations is measured, and then the angles between these two stations and a third station are measured. The Law of Sines is then used to calculate the lengths of the other two sides of the triangle created by the three stations. The advantage of the Law of Sines is that it is generally easier to measure angles than it is to measure distances, especially if the terrain is rough. We can use the Law of Sines to find the height of a mountain or the elevation of an airplane. The highest peak of the Himalayas, Mt. Everest, was measured using the Law of Sines, and its height was recorded as 29,002 ft. Satellite measurements of the height of Mt. Everest record its height as roughly 29,028 ft. These two measurements differ by less than 0.1 percent! The next example illustrates one of these types of applications of the Law of Sines.

Example 6

To measure the height of a mountain, a surveyor takes two measurements of the angle of inclination of the peak of the mountain. The first measurement of the angle of inclination is 35°. The surveyor moves 2,500 feet, and the angle of elevation is 40°. If the survey equipment is 4 feet high, find the height of the mountain.

Solution: The key to working out application problems is to look at the world through triangular eyes. In other words, look for various triangles, and see if either the Pythagorean Theorem or the Law of Sines can be used. Figure 6.8 will help illustrate the situation presented in this example.

First, consider the obtuse triangle formed by the points A, B, and C. The interior angles of the triangle measure 35°, 140°, and 5°. We also know the length of one side of the triangle: 2,500 feet. We can use the Law of Sines:

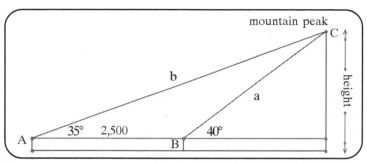

Figure 6.8.

$$\frac{\sin\alpha}{a}=\frac{\sin\beta}{b}=\frac{\sin\gamma}{c}$$

$$\frac{\sin 35°}{a}=\frac{\sin 140°}{b}=\frac{\sin 5°}{2500}$$

We can solve for either a or b, and then use the appropriate right triangle, either $\triangle BCD$ if we solve for a or $\triangle ACD$ if we solve for b, to find the height of the peak. I will solve for a:

$$\frac{\sin 35°}{a}=\frac{\sin 5°}{2500}$$

$$a=\frac{2500\sin 35°}{\sin 5°}\approx 16453$$

Now, solve for the height using the definition of the sine ratio applied to $\triangle BCD$. Be sure to include the height of the equipment in your calculations:

$$\sin 40°=\frac{\text{height}-4}{a}$$

height $= a \sin 40° + 4$

height $= 16453 \sin 40° + 4 \approx 10580$

The peak of the mountain is approximately 10,580 feet.

The application problems can get a bit tricky. Remember to draw lots of triangles. Keep in mind that we are no longer limited to working with right triangles. When working with the Pythagorean Theorem, you must have a right triangle. One advantage of the Law of Sines is that it applies to triangles in general!

Lesson 6-2 Review

1. Find the measure of the other angle and the length of the other two sides of the triangle shown in Figure 6.9.

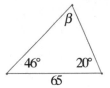

Figure 6.9.

2. Find the measures of the other two angles and the length of the other side of the triangle shown in Figure 6.10.

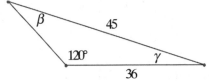

Figure 6.10.

3. A pilot is flying over a straight runway. There are two markers, 5 miles apart. The angles of depression of these markers are 32° and 48°, as shown in Figure 6.11. Find the elevation of the plane.

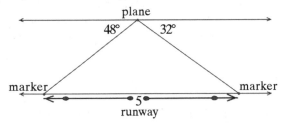

Figure 6.11.

Lesson 6-3: The Law of Cosines

We can use the Pythagorean Theorem to calculate lengths and angles of right triangles. For right triangles, we need to know either the length of two sides, or the length of one side and the measure of one of the acute angles of the triangle. Remember that with a right triangle, we already know the measure of one of the angles: 90°. Taken together, we actually know three of the six measurable quantities of a triangle, and with the Pythagorean Theorem we can find the other three quantities.

The Law of Sines enables us to calculate lengths and angles of triangles in general. If we know the length of two sides of a triangle and the measure of one of the angles *opposite* one of the sides, or if we know the length of one side and the measure of two angles, we can find the other three quantities. The Law of Sines will not apply if we only know the lengths of three sides, or if we only know the length of two sides and the measure of the *included* angle. In these situations, we need the *Law of Cosines*.

Given a triangle with sides of length a, b, and c, and opposite angles α, β, and γ, the **Law of Cosines** consists of three equations:

$$c^2 = a^2 + b^2 - 2ab \cos \gamma$$

$$b^2 = a^2 + c^2 - 2ac \cos \beta$$

$$a^2 = b^2 + c^2 - 2bc \cos \alpha$$

The Pythagorean Theorem follows from the Law of Cosines. If γ is the right angle of a triangle, and c is the length of the hypotenuse, then using the Law of Cosines, we have:

$$c^2 = a^2 + b^2 - 2ab \cos \frac{\pi}{2} = a^2 + b^2$$

In other words, the Pythagorean Theorem is a special case of the Law of Cosines. The Law of cosines can be used with *all* triangle, whereas the Pythagorean Theorem can only be used with *right* triangles.

The Law of Cosines can be used to find all three angles if the lengths of all three sides are known. Also, if the length of two sides and the measure of the *included* angle are known, then the measure of the other two angles and the length of the third side can be found using the Law of Cosines. Remember that if the length of two sides and the measure of one of the opposite angles are known, then the Law of Sines can be used to find the measure of the other two angles and the length of the third side. In this situation, it is possible that there are zero, one, or two triangles that satisfy the given information. With the Law of Cosines, if there is a solution, it will be unique. It is possible that no triangle satisfies the given conditions.

Example 1

The lengths of the sides of a triangle are $a = 5$, $b = 8$, and $c = 12$. Classify the triangle as acute, right, or obtuse.

Solution: In order to classify this triangle as acute, right, or obtuse, we need to find the measures of the three angles. To find the measures of the angles, we will use all three equations in the Law of Cosines:

$c^2 = a^2 + b^2 - 2ab \cos \gamma$	$b^2 = a^2 + c^2 - 2ac \cos \beta$	$a^2 = b^2 + c^2 - 2bc \cos \alpha$
$12^2 = 5^2 + 8^2 - 2(5)(8)\cos \gamma$	$8^2 = 5^2 + 12^2 - 2(5)(12)\cos \beta$	$5^2 = 8^2 + 12^2 - 2(8)(12)\cos \alpha$
$80 \cos \gamma = -55$	$120 \cos \beta = 105$	$192 \cos \alpha = 183$
$\cos \gamma = -0.688$	$\cos \beta = 0.875$	$\cos \alpha = 0.953$
$\gamma = 133.4°$	$\beta = 29.0°$	$\alpha = 17.6°$

We can check our answer by adding the three angles together. Their sum should be 180°: $133.4° + 29.0° + 17.6° = 180°$. Looking at the measures of the three angles, this triangle is obtuse.

In Example 1, we could have just used the Law of Cosines once and then used the Law of Sines once we knew the measure of one angle. We could have also used the Law of Cosines twice, and then used the fact that the interior angles of a triangle must add up to 180° to find the third angle. If we use the Law of Cosines three times, then we have a way to check our answers. If you can apply the Law of Cosines once, then you can certainly apply it twice or three times. The ability to check your work is certainly worth using the Law of Cosines a few more times.

Example 2

For the triangle shown in Figure 6.12, find the measure of the third side and the measures of the other two angles.

Solution: Start by using the third equation of the Law of Cosines to find a:

$a^2 = b^2 + c^2 - 2bc \cos \alpha$

$a^2 = 25^2 + 15^2 - 2(25)(15)\cos 40° \approx 275.5$

$a = 16.6$

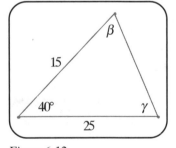

Figure 6.12.

Now we can use the other two Law of Cosine equations to find the measures of the other two angles:

$c^2 = a^2 + b^2 - 2ab \cos \gamma$ $b^2 = a^2 + c^2 - 2ac \cos \beta$

$15^2 = 16.6^2 + 25^2 - 2(16.6)(25)\cos \gamma$ $25^2 = 16.6^2 + 15^2 - 2(16.6)(15)\cos \beta$

$830 \cos \gamma = 675.6$ $498 \cos \beta = -124.44$

$\cos \gamma = 0.814$ $\cos \beta = -0.25$

$\gamma = 35.5°$ $\beta = 104.5°$

As a final check, we can verify that the measures of the interior angles of this triangle add up to 180°: 40° + 104.5° + 35.5° = 180°

To find the measures of the other two angles, we could have used the Law of Sines as well. One advantage to using the Law of Cosines instead of the Law of Sines is that the sign of the cosine function enables us to distinguish between an acute angle and an obtuse angle. Remember that the cosine of an obtuse angle is negative. When we use the Law of Sines to find an angle in a triangle, we are actually calculating the value of the *reference* angle. The sine of an angle is positive if the angle lies in Quadrant I or Quadrant II. In other words, the sine of an *obtuse* angle is positive and has the same value as the sine of its corresponding reference angle. We must examine both angles when we work with the Law of Sines. The Law of Cosines involves the cosine function, and the sign of the cosine of an angle in a triangle reveals whether the angle is acute or obtuse.

Example 3

A golf ball is 40 yards from a marker that is in the center of the fairway. The marker is 120 yards from the center of the green. The angle between the center of the green and the ball is 100°, as shown in Figure 6.13. The caddie reports this information to the golfer. How far is the ball from the center of the green?

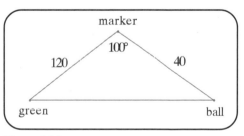

Figure 6.13.

Solution: We can use the Law of Cosines to find the distance between the ball and the center of the green. Let $a = 120$, $b = 40$, and $\gamma = 100°$. Using the Law of Cosines, we have:

$c^2 = a^2 + b^2 - 2ab \cos \gamma$

$c^2 = 120^2 + 40^2 - 2(120)(40)\cos 100°$

$c^2 = 17667$

$c = 133$

The ball is approximately 133 yards from the center of the green.

---

1. The lengths of the sides of a triangle are $a = 7$, $b = 10$, and $c = 13$. Classify the triangle as acute, right, or obtuse.

2. For the triangle shown in Figure 6.14, find the measure of the third side and the measures of the other two angles ($\alpha = 55°$, $b = 10$, and $c = 14$).

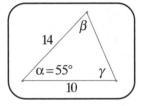

3. A golf ball is 60 yards from a marker that is in the center of the fairway. The marker is 150 yards from the center of the green. The

Figure 6.14.

angle between the center of the green and the ball is 115°. The caddie reports this information to the golfer. How far is the ball from the center of the green?

Lesson 6-4: Area of a Triangle

The most common formula for finding the area of a triangle is $A = \frac{1}{2}bh$, where b represents the length of the base of the triangle and h represents the corresponding height of the triangle. It does not matter which side of the triangle serves as the base. The height of the triangle will change accordingly, so that the quantity $\frac{1}{2}bh$ remains the same. The line segment used to calculate the height of the triangle must be perpendicular to the base of the triangle. In order to measure the area of a triangle using the formula $A = \frac{1}{2}bh$, we must construct a right triangle. Where there is a right triangle, there are trigonometric ratios just waiting to be used.

If $\triangle ABC$ is a general triangle, as show in Figure 6.15, we can use the trigonometric ratios to explore the relationship between the height of the triangle and the sides of the triangle. Remember that if we have a right triangle, we can make use of the Pythagorean Theorem to find the length of one side if we know the lengths of the other two sides. We can also use the trigonometric ratios if we know the measure of an angle and the length of a side. For triangles in general, we can use the Law of Sines or the Law of Cosines to solve for either lengths of sides or measures of angles.

Example 1

Find the area of the triangle shown in Figure 6.15.

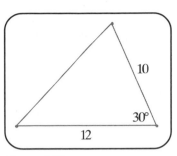

Figure 6.15.

Solution: Let b represent the base of the triangle. The height of the triangle can be found by creating a right triangle and using the definition of the sine ratio:

$$h = 10\sin 30° = 10\left(\tfrac{1}{2}\right) = 5$$

The area of the triangle is:

$$A = \tfrac{1}{2}bh$$

$$A = \tfrac{1}{2}(12)(5) = 30$$

In general, if you are given the length of two sides of a triangle and the measure of the included angle, the sine function will be the key to finding the area. For the triangle shown in Figure 6.16, the height of the triangle can be found using the definition of the sine ratio: $h = a \sin \gamma$. The length of the base is b, so the formula for the area of the triangle can be written as:

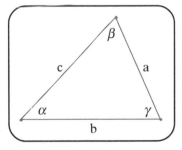

Figure 6.16.

$$A = \tfrac{1}{2}ab\sin\gamma$$

There is another formula that can be used to calculate the area of a triangle that bypasses the tedious process of finding the height of the triangle. This formula, known as *Heron's Formula*, has been around for thousands of years and only involves the lengths of the sides of the triangle. To calculate the area of a triangle, first calculate the **semi-perimeter**, s, which is defined to be one-half the perimeter of the triangle: $s = \tfrac{1}{2}(a+b+c)$. **Heron's Formula** is as follows:

$$A = \sqrt{s(s-a)(s-b)(s-c)}$$

In general, given three pieces of information about a triangle, including the length of at lease one side of the triangle, it is possible to either find the lengths of two sides and the included angle, or to find the lengths of all three sides of the triangle. From there, you can either create a right triangle and find the area using the formula $A = \frac{1}{2}ab\sin\gamma$, or use Heron's Formula, $A = \sqrt{s(s-a)(s-b)(s-c)}$, to find the area of the triangle.

Example 2

Find the area of a triangle whose sides have lengths $a = 10$, $b = 12$, and $c = 18$.

Solution: Because the lengths of all three sides are given, we should use Heron's Formula. First, calculate the semi-perimeter:

$$s = \frac{1}{2}(10 + 12 + 18) = 20$$

Next, use Heron's Formula:

$$A = \sqrt{s(s-a)(s-b)(s-c)}$$

$$A = \sqrt{20(20-10)(20-12)(20-18)}$$

$$A = \sqrt{20 \cdot 10 \cdot 8 \cdot 2} = 40\sqrt{2}$$

Example 3

Three circles of radii 3, 4, and 5 cm are mutually tangent, as shown in Figure 6.17. Find the area enclosed between the three circles.

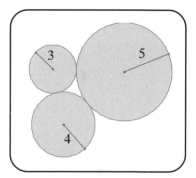

Solution: This problem involves circles rather than triangles, but we can construct a triangle using the centers of the three circles, as shown in Figure 6.18. The area of the shaded region will be the difference between the area of *Figure 6.17.* the triangle and the areas of the three sectors of the circles. In order to find the area of the triangle, we will use Heron's Formula. We will also need to know the areas of

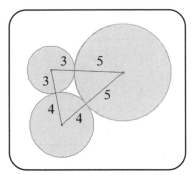

Figure 6.18.

the sectors, which means that we will need to know the three angles of the triangle. We will use the Law of Cosines to find the three angles. In the applications we have seen in this chapter, the units for the angle measures have been degrees. Remember that in the formula for calculating the area of a sector, $A = \frac{1}{2}r^2\theta$, the angle θ is measured in *radians*. We need to be consistent with our sides and corresponding angles, because the area of the sector of a circle depends on the central angle of the sector as well as the radius of the circle. Figure 6.18 shows the circles, the sides of the triangle, and the corresponding angles.

First, we will find the lengths of the three sides of the triangle, the semi-perimeter, and then the area of the triangle using Heron's Formula. The lengths of the sides of the triangle can be found by using the radii of the circles. The lengths of the sides of the triangle are: $a = 7$, $b = 8$, and $c = 9$. The semi-perimeter is $s = \frac{1}{2}(7+8+9) = 12$. Using Heron's Formula, we have:

$$A = \sqrt{s(s-a)(s-b)(s-c)}$$

$$A = \sqrt{12(12-7)(12-8)(12-9)}$$

$$A = \sqrt{12 \cdot 5 \cdot 4 \cdot 3} = 12\sqrt{5} \approx 26.83$$

Next, we will find the three angles of the triangle using the Law of Cosines. The units for the angle measures will be in radians.

$c^2 = a^2 + b^2 - 2ab \cos \gamma$	$b^2 = a^2 + c^2 - 2ac \cos \beta$	$a^2 = b^2 + c^2 - 2bc \cos \alpha$
$9^2 = 7^2 + 8^2 - 2(7)(8)\cos \gamma$	$8^2 = 7^2 + 9^2 - 2(7)(9)\cos \beta$	$7^2 = 8^2 + 9^2 - 2(8)(9)\cos \alpha$
$112 \cos \gamma = 32$	$126 \cos \beta = 66$	$144 \cos \alpha = 96$
$\cos \gamma = 0.286$	$\cos \beta = 0.524$	$\cos \alpha = 0.667$
$\gamma = 1.28$	$\beta = 1.02$	$\alpha = 0.841$

The areas of the sectors are:

$A_1 = \frac{1}{2}3^2\gamma$	$A_2 = \frac{1}{2}4^2\beta$	$A_3 = \frac{1}{2}5^2\alpha$
$A_1 = \frac{1}{2}3^2(1.28)$	$A_2 = \frac{1}{2}4^2(1.02)$	$A_3 = \frac{1}{2}5^2(0.841)$
$A_1 = 5.76$	$A_2 = 8.16$	$A_3 = 9.26$

The total area of the sectors of the circles is:

$A_1 + A_2 + A_3 = 5.76 + 8.16 + 9.26 = 23.18$

The area of the shaded region is the difference between the area of the triangle, which is 26.83, and the area of the sectors of the circles, which is 23.18. The area of the shaded region is:

$A_{shaded} = 26.83 - 23.18 = 3.65$

Lesson 6-4 Review

1. Find the area of the triangle shown in Figure 6.19.

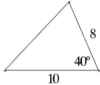

Figure 6.19.

2. Find the area of the triangle whose sides have length $a = 6$, $b = 12$, and $c = 15$.

Answer Key

Lesson 6-1 Review

1. Construct a right triangle with the ladder as the hypotenuse, the ground as the base, and the wall as the vertical leg. Then the length of the base is 10 ft. and the length of the hypotenuse is 40 ft. The angle formed by the ladder and the *building* is:

$\beta = \sin^{-1}\frac{10}{40} \approx 14.5°$

2. Create a right triangle where the tree forms the vertical leg and the shadow forms the horizontal leg. The angle formed by the hypotenuse and the ground is the angle of elevation, and the height of the tree is $500 \tan 23.6° \approx 218.4$. The tree is 218.4 ft. high.

3. Figure 6.20 will help visualize the situation. Construct two right triangles. The building serves as the vertical leg of the first triangle, and the building and antenna serve as the vertical leg of the second triangle. The horizontal leg of both triangles is 300 ft.

    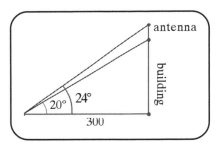

    Figure 6.20.

    The length of the building is $300 \tan 20°$ ≈ 109.2, and the length of the building and the antenna is $300 \tan 241° \approx 133.6$. From these two lengths, we see that the length of the antenna is 24.4 ft.

Lesson 6-2 Review

1. The measure of the third angle is $180° - 46° - 20° = 144°$.

    Using the Law of Sines, we have:

    $$\frac{\sin 46°}{a} = \frac{\sin 114°}{65} = \frac{\sin 20°}{c}, \text{ so } a = \frac{65 \sin 46°}{\sin 114°} \approx 51.2 \text{ and } b = \frac{65 \sin 20°}{\sin 114°} \approx 24.3.$$

2. Using the Law of Sines, we have:

    $$\frac{\sin 120°}{45} = \frac{\sin \beta}{36} = \frac{\sin \gamma}{c}, \text{ so } \sin \beta = \frac{36 \sin 120°}{45} \approx 0.693, \text{ so } b \gg 43.9° \text{ or } b \gg 136.1°.$$

    If $\beta \approx 43.9°$, then $\gamma = 180° - 120° - 43.9° = 16.1°$,

    and $\dfrac{\sin 120°}{45} = \dfrac{\sin 16.1°}{c}$, so $c = \dfrac{45 \sin 16.1°}{\sin 120°} \approx 14.4$.

    There is no solution if $\beta \approx 136.1°$.

3. The measure of the angle that the plane makes with the two markers is $180° - 48° - 32° = 100°$. Let x represent the distance between the plane and the marker to the left.

    Then $\dfrac{\sin 100°}{5} = \dfrac{\sin 32°}{x}$, and $x = \dfrac{5 \sin 32°}{\sin 100°} \approx 2.69$.

    The height of the triangle formed by the plane and the two markers is given by height $= x \sin 48° = (2.69) \sin 48° \approx 2.00$.

    The altitude of the plane is approximately 2 miles.

Lesson 6-3 Review

1. $c^2 = a^2 + b^2 - 2ab \cos \gamma$

   $13^2 = 7^2 + 10^2 - 2 \cdot 7 \cdot 10 \cos \gamma$

   $140 \cos \gamma = -20$

   $\cos \gamma = -0.143$

   $\gamma = 98.2°$

   Because the measure of γ is greater than 90°, the triangle is obtuse.

2. First, find a:

   $a^2 = b^2 + c^2 - 2bc \cos \alpha$

   $a^2 = 10^2 + 14^2 - 2(10)(14)\cos 55°$

   $a^2 = 135.39$

   $a = 11.6$

   Use a to find the measure of β:

   $b^2 = a^2 + c^2 - 2ab \cos \beta$

   $10^2 = 11.6^2 + 14^2 - 2(11.6)(14) \cos \beta$

   $324.8 \cos \beta = 230.6$

   $\beta \approx 44.8°$

   Finally, find the measure of γ. $\gamma = 180° - 55° - 44.8° = 80.2°$

3. Set up a triangle and apply the Law of Cosines:

   $a^2 = b^2 + c^2 - 2bc \cos \alpha$

   $a^2 = 60^2 + 150^2 - 2(60)(150)\cos 115°$

   $a^2 \approx 33,707$

   $a = 183.6$

   The golf ball is approximately 183.6 yards from the center of the green.

Lesson 6-4 Review

1. area $= \frac{1}{2}(12)(8 \sin 40°) \approx 30.9$

2. $s = \frac{1}{2}(15 + 6 + 12) = 16.5$, and

   area $= \sqrt{16.5(16.5 - 6)(16.5 - 12)(16.5 - 16)} = \sqrt{16.5 \cdot 10.5 \cdot 4.5 \cdot 1.5} \approx 34.2$

Polar Coordinates

The Cartesian coordinate system is used to uniquely describe the location of any point in the plane. The Cartesian coordinate system is also called the rectangular coordinate system. In the rectangular coordinate system, the location of a point is specified by an ordered pair of the form (x, y), where x represents the perpendicular distance between the point and the y-axis, and y represents the perpendicular distance between the point and the x-axis. The rectangular coordinate system can be viewed as a grid of rectangles.

There are other coordinate systems that can be used to describe the location of any point in the plane, such as the polar coordinate system. The polar coordinate system can be viewed as a grid of concentric circles centered at the origin. There are advantages to using a polar coordinate plane rather than a rectangular coordinate plane. If the geometry of a curve is circular, the polar coordinate plane may be better suited to describe its shape. On the other hand, when working with geometric shapes that involve line segments, parabolas, or polynomial functions, the rectangular coordinate system is the one to use.

Lesson 7-1: Polar Coordinates

A point in the plane can be described by the perpendicular distances between the point and each of the coordinate axes. In addition, a point in the plane can be specified by the distance from the point to the origin, and the angle that the ray starting at the origin and passing through the point makes with the positive x-axis. In the polar coordinate system, the location of a point is specified by an ordered pair of the form (r, θ), where r represents the distance between the point and the origin, and θ

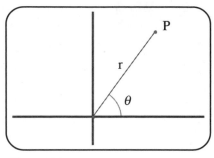

Figure 7.1.

represents the angle formed by the ray starting at the origin and passing through the point and the positive *x*-axis, as shown in Figure 7.1.

The distance between the point and the origin is called the **radial distance**. The distance between two points is always a positive number, so one of the early restrictions on *r* is that r > 0. It is possible to relax this restriction on *r*, as we will see later.

The rectangular coordinates of the point *P*, as shown in Figure 7.2, is $\left(\sqrt{2},\sqrt{2}\right)$. The distance between *P* and the origin can be found using the Pythagorean Theorem:

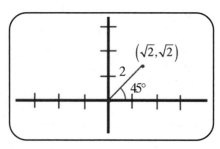

Figure 7.2.

$$r=\sqrt{\left(\sqrt{2}\right)^2+\left(\sqrt{2}\right)^2}=\sqrt{2+2}=\sqrt{4}=2$$

The angle that the ray $\overrightarrow{OP}$ makes with the *x*-axis is 45°, so the ordered pair (2, 45°) would be a polar coordinate representation of the point whose rectangular coordinates are $\left(\sqrt{2},\sqrt{2}\right)$.

In Chapter 3, we learned how to interpret angles whose measure is negative or greater than 360°. It is difficult to distinguish between a 45° angle and a –315° angle, because the terminal side of these two angles coincide. One of the disadvantages to the polar coordinate system is that there is not a unique ordered pair that describes the location of a point in the plane. For example, the ordered pairs (2, 45°) and (2, –315°) can both be used to represent a point whose rectangular coordinates are $\left(\sqrt{2},\sqrt{2}\right)$.

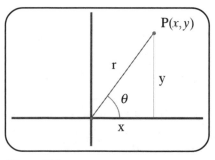

Figure 7.3.

Figure 7.2 should have reminded you of the methods we used to define the trigonometric ratios. In fact, the trigonometric functions can be used to establish a relationship between the

Cartesian coordinates of a point and the polar coordinates of that point. If the Cartesian coordinates of a point P are (x, y), as shown in Figure 7.3, then the corresponding polar coordinates of that point can be found using the Pythagorean Theorem and the definition of the tangent function:

$r^2 = x^2 + y^2$ and $\tan\theta = \frac{y}{x}$, as long as $x \neq 0$

We can use these two equations to solve for r and θ:

$r = \sqrt{x^2 + y^2}$ and $\theta = \tan^{-1}\frac{y}{x}$

We can also convert between the polar coordinates of a point and its corresponding rectangular coordinates. If the polar coordinates of a point P are (r, θ), then using the definition of the sine and cosine ratios, we can find the x- and y-coordinates of the point:

$\cos\theta = \frac{x}{r}$ and $\sin\theta = \frac{y}{r}$

From this, we see:

$x = r \cos \theta$ and $y = r \sin \theta$

So, given the coordinates of a point using one system, we can easily convert to coordinates in the other. Keep in mind that the polar coordinate representation of a point is not unique.

Example 1

Find the rectangular coordinates for the point that has polar coordinates $\left(5, \frac{3\pi}{4}\right)$.

Solution: The point lies in Quadrant II, because $\frac{\pi}{2} < \theta < \pi$. Use the equations $x = r \cos \theta$ and $y = r \sin \theta$ to solve for x and y:

$x = 5\cos\frac{3\pi}{4} = -\frac{5\sqrt{2}}{2}$

$y = 5\sin\frac{3\pi}{4} = \frac{5\sqrt{2}}{2}$

The rectangular coordinates of the point are $\left(-\frac{5\sqrt{2}}{2}, \frac{5\sqrt{2}}{2}\right)$.

Example 2

Find a polar coordinate representation for the point that has rectangular coordinates $\left(-1, -\sqrt{3}\right)$.

Solution: This point lies in Quadrant III. Use the equations

$r = \sqrt{x^2 + y^2}$ and $\theta = \tan^{-1}\frac{y}{x}$ to solve for r and θ:

$r = \sqrt{(-1)^2 + (-\sqrt{3})^2} = \sqrt{1+3} = \sqrt{4} = 2$

$\theta = \tan^{-1}\left(\frac{-\sqrt{3}}{-1}\right) = \tan^{-1}\left(\sqrt{3}\right) = \frac{\pi}{3}$

Because of the restriction on the domain and range of the inverse tangent function, this angle is the reference angle.

The point $\left(-1, -\sqrt{3}\right)$ lies in Quadrant III, so $\theta = \frac{4\pi}{3}$.

The point $\left(-1, -\sqrt{3}\right)$ in polar coordinates is written $\left(2, \frac{4\pi}{3}\right)$.

When converting between coordinate systems, it will be helpful to be able to recognize when you are working with the special angles that were discussed in Chapter 2.

The formulas $x = r \cos\theta$ and $y = r \sin\theta$ will enable us to convert equations from rectangular coordinates to polar coordinates quite easily. The parabola $y = 2x^2$ can be written using polar coordinates by substituting for x and y, and then solving for r in terms of trigonometric functions evaluated at θ:

Start with the equation in rectangular coordinates

$$y = 2x^2$$

Substitute in for x and y $\qquad\qquad r \sin\theta = 2(r \cos\theta)^2$

Expand $(r \cos\theta)^2$ $\qquad\qquad r \sin\theta = 2r^2 \cos^2\theta$

Divide by $2r \cos^2\theta$ and cancel common terms in each fraction

$$\frac{r \sin\theta}{2r \cos^2\theta} = \frac{2r^2 \cos^2\theta}{2r \cos^2\theta}$$

$$r = \frac{\sin\theta}{2 \cos^2\theta}$$

Use the definition of the tangent function and the secant function to simplify $\qquad\qquad\qquad\qquad r = \frac{1}{2}\tan\theta \sec\theta$

Your experience proving various trigonometric identities should convince you that there are many equivalent forms for this equation. I have written it as $r = \frac{1}{2}\tan\theta \sec\theta$, but I could have just as easily written it as $r = \frac{1}{2}\sin\theta \sec^2\theta$.

Converting equations from polar coordinates to rectangular coordinates often requires more work. In Chapter 9, we will derive polar equations for some of the more interesting geometrical shapes, like ellipses, parabolas, and hyperbolas. For now, you should be familiar with the standard formula for a line:

$$Ax + By = C$$

and a circle with center (h, k) and radius r:

$$(x - h)^2 + (y - k)^2 = r^2$$

We will establish the polar coordinate versions of these two types of functions.

Example 3

Express the following polar equations in rectangular coordinates and briefly describe their graphs:

a. $r = 2 \csc \theta$ b. $r = 4 \cos \theta$

Solution: We will need to make use of the equations $x = r \cos \theta$, $y = r \sin \theta$, $r^2 = x^2 + y^2$, and $\tan\theta = \frac{y}{x}$. If the equation we are starting with involves sine or cosine functions, try to combine them with an r to turn them into x or y. Sometimes it helps to multiply both sides of an equation by r; other times it helps to multiply both sides of an equation by $\cos \theta$ or $\sin \theta$. There are times when the identity $\cos^2 \theta + \sin^2 \theta = 1$ can be used to eliminate θ in an equation.

a. $r = 2 \csc \theta$

Rewrite the equation in terms of the sine function $r = \dfrac{2}{\sin\theta}$

Multiply both sides of the equation by $\sin \theta$ $r \sin \theta = 2$

Use the fact that $y = r \sin \theta$ to eliminate the polar variables

$$y = 2$$

This is an equation for a horizontal line.

b. $r = 4 \cos \theta$

Multiply both sides of the equation by r $r^2 = 4r \cos \theta$

Use the equations $r^2 = x^2 + y^2$ and $x = r \cos \theta$ to eliminate the polar variables

$$x^2 + y^2 = 4x$$

Subtract $4x$ from both sides of the equation $x^2 - 4x + y^2 = 0$

Complete the square in x $(x - 2)^2 + y^2 = 4$

This is a circle with center $(2, 0)$ and radius 2

Lesson 7-1 Review

1. Find the rectangular coordinates for the point that has polar coordinates $\left(2, \frac{2\pi}{3}\right)$.

2. Find a polar coordinate representation for the point that has rectangular coordinates $(-3, 3)$.

3. Express the following polar equations in terms of rectangular coordinates:

    a. $r = 3$ b. $\sec \theta = 2$

4. Convert the function $y = x$ to polar form.

Lesson 7-2: Graphs of Functions in Polar Coordinates

The first technique that you used to graph functions in rectangular coordinates was to evaluate the function for various values of the independent variable and then plot the resulting points, or ordered pairs. To graph a polar equation, we will start with the same procedure. We will begin by graphing polar equations of the form $r = f(\theta)$. When plotting points that are given in rectangular coordinates, we divide the plane into a rectangular grid. When plotting points that are

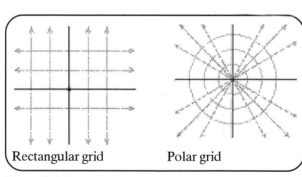

Rectangular grid   Polar grid

Figure 7.4.

given in polar coordinates, we will need to divide the plane into regions that take advantage of the benefits of the polar coordinate system. The location of a point, P, in polar coordinates is specified by its radial distance to the origin and the angle that the ray $\overrightarrow{OP}$ makes with the positive x-axis. The positive x-axis is called the **polar axis**. To graph points using polar coordinates, we will divide the plane into concentric circles centered at the origin, with the special angles that we discussed in Chapter 2 marked off, as shown in Figure 7.4.

Each circle in the polar coordinate grid represents the set of points that are a fixed, or constant, distance from the origin. Functions of the form $r = c$, where c is a constant, can be graphed easily using this polar grid: they are circles centered at the origin with radius c. To represent the function $r = c$ in rectangular coordinates, square both sides of the equation: $r^2 = c^2$. Now use the relationship $x^2 + y^2 = r^2$ to eliminate r^2 from the equation: $x^2 + y^2 = c^2$. We recognize this as the equation of a circle centered at the origin with radius c. The graph of the circle $r = 2$ is shown in Figure 7.5.

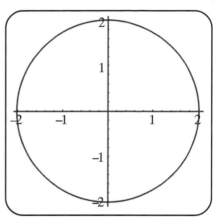

Figure 7.5.

Functions of the form $\theta = c$, where c is a constant, can also be graphed easily using this polar grid: they are rays starting at the origin that make an angle θ with the positive x-axis. In fact, the positive x-axis can be represented by the function $\theta = 0$. To represent the function $\theta = c$ in rectangular coordinates, apply the tangent function to both sides: $\tan \theta = \tan c$. Now use the relationship $\tan \theta = \frac{y}{x}$ to eliminate $\tan \theta$ from the equation: $\frac{y}{x} = \tan c$.

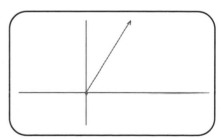

Figure 7.6.

Multiply both sides of the equation by x: $y = (\tan c)x$. This is the equation of a line passing through the origin with slope $\tan c$. The graph of $\theta = c$ will only be half of the line $y = (\tan c)x$. The graph of $\theta = \frac{\pi}{3}$ is shown in Figure 7.6.

In general, graphs of polar equations can be more complicated than $r = c$ or $\theta = c$. In order to graph a more complicated polar equation of the form $r = f(\theta)$, I recommend evaluating the equation for various values of θ and plotting the resulting points. The values of θ at which you choose to evaluate the function should be related to the special angles that were discussed in Chapter 2. Those special angles have been used throughout this book. Let me emphasize that it is extremely important to be familiar with the values of the trigonometric functions at those special angles. Your ability to graph polar equations will depend on it.

When graphing polar equations, we can use the periodic nature of the trigonometric functions to help determine the range of values of θ to use. Once we graph one cycle of a function, we will have its entire graph. A cycle is complete when $f(\theta + 2n\pi) = f(\theta)$, for some positive integer n. To find the range of angles, we will need to find n.

Another thing to take into consideration when graphing a polar equation is symmetry. If a polar equation satisfies the equation $f(-\theta) = f(\theta)$, then its graph will be symmetric about the x-axis. If a polar equation satisfies the equation $f(\pi - \theta) = f(\theta)$, then the graph will be symmetric about the y-axis. If a polar equation remains the same when r is replaced by $-r$, then its graph is symmetric with respect to the origin.

When we first introduced polar coordinates, I mentioned that r represents the distance between a point and the origin, and that r must be positive. When evaluating polar functions, we will interpret a point $(-r, \theta)$ as equivalent to the point $(r, \theta - \pi)$. The points (r, θ) and $(-r, \theta)$ represent collinear points that are the same distance from the origin. For ex-

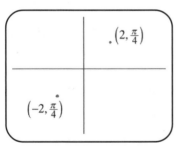

Figure 7.7.

ample, the points $\left(-2, \frac{\pi}{4}\right)$ and $\left(2, \frac{\pi}{4}\right)$ are shown in Figure 7.7. Notice that the points $\left(-2, \frac{\pi}{4}\right)$ and $\left(2, -\frac{3\pi}{4}\right)$ are equivalent to each other.

The following table may help you visualize the effects that a negative value of r has on a point in the various quadrants.

Sign of r	Quadrant of θ	Quadrant of (r, θ)
–	I	III
–	II	IV
–	III	I
–	IV	II

Example 1

Sketch the graph of $r = 3\sin \theta$.

Solution: We only need to evaluate this function for angles that lie between 0 and π, or for angles in Quadrant I and Quadrant II. For angles in Quadrant III and Quadrant IV, the sine will be negative, and r will be negative. From the previous table, we see that a negative value for r with a corresponding angle in Quadrant III or Quadrant IV will be equivalent to a point whose r-coordinate is positive and whose θ-coordinate is in Quadrant I or Quadrant II. So evaluating the function for angles between π and 2π will only duplicate the work done in evaluating the function for angles between 0 and π. If you need to be convinced of this fact, you should evaluate the function for the special angles between π and 2π and see for

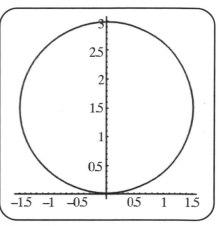

Figure 7.8.

yourself. The table below shows the values of this function for the special angles that lie between 0 and π.

θ	0	$\frac{\pi}{6}$	$\frac{\pi}{4}$	$\frac{\pi}{3}$	$\frac{\pi}{2}$	$\frac{2\pi}{3}$	$\frac{3\pi}{4}$	$\frac{5\pi}{6}$	π
$3\sin \theta$	0	$\frac{3}{2}$	$\frac{3\sqrt{2}}{2}$	$\frac{3\sqrt{3}}{2}$	3	$\frac{3\sqrt{3}}{2}$	$\frac{3\sqrt{2}}{2}$	$\frac{3}{2}$	0

The graph of this function is shown in Figure 7.8.

Example 2

Sketch the graph of $r = \sin 2\theta$.

Solution: The symmetry of the sine functions suggests that the graph of $r = \sin 2\theta$ should be symmetrical, but the specific symmetry is not obvious. We can think about the shape of this curve, and then plot the points and graph it. It is sometimes helpful to graph the polar function using rectangular coordinates (meaning that the horizontal axis represents θ and the vertical axis represents r) and visualizing how r changes with θ. That information can then be transferred to the polar grid and a sketch of the graph can be made. The graph of $r = \sin 2\theta$ as θ ranges from 0 to 2π from a rectangular perspective is shown in Figure 7.9.

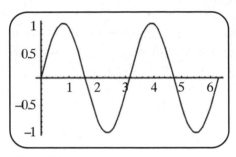

Figure 7.9.

When $\theta = 0, r = 0$, so the graph starts out at the origin. As θ increases from 0 to $\frac{\pi}{4}$, $\sin 2\theta$ increases from 0 to 1. As θ increases from $\frac{\pi}{4}$ to $\frac{\pi}{2}$, $\sin 2\theta$ then decreases back to 0. The graph of this first part is shown in Figure 7.10.

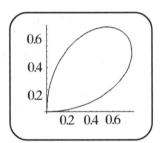

Figure 7.10.

We will now consider angles that are in Quadrant II. As θ increases from $\frac{\pi}{2}$ to $\frac{3\pi}{4}$, $\sin 2\theta$ decreases from 0 to -1. As θ increases from $\frac{3\pi}{4}$ to π, $\sin 2\theta$ increases from -1 to 0. Because r is negative and θ is in Quadrant II, the next stage of the graph will actually be drawn in Quadrant IV. It will look very similar to the first curve that was drawn, and it is shown in Figure 7.11.

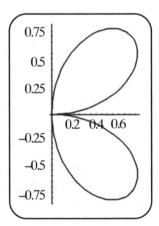

Figure 7.11.

We will now consider angles that are in Quadrant III. As θ increases from π to $\frac{5\pi}{4}$, $\sin 2\theta$ increases from 0 to 1. As θ increases from $\frac{5\pi}{4}$ to $\frac{3\pi}{2}$, $\sin 2\theta$ decreases from 1 to 0. Because r is positive and θ is in Quadrant III, the next stage of the graph will be drawn in Quadrant III. It will look very similar to the first two curves that were drawn, and it is shown in Figure 7.12.

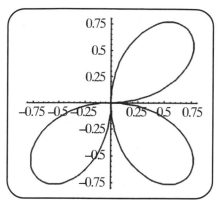

Figure 7.12.

Finally, we will consider angles that are in Quadrant IV. As θ increases from $\frac{3\pi}{2}$ to $\frac{7\pi}{4}$, sin 2θ decreases from 0 to −1. As θ increases from $\frac{7\pi}{4}$ to 2π, sin 2θ increases from −1 to 0. Because r is negative and θ is in Quadrant IV, the next stage of the graph will actually be drawn in Quadrant II. It will look very similar to the first three curves that were drawn, and it is shown in Figure 7.13.

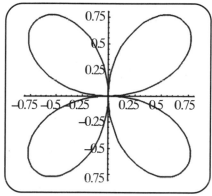

Figure 7.13.

The graph of $r = \sin 2\theta$ is called a *four-leaf rose*. In general, the graph of an equation of the form $r = a \sin n\theta$ or $r = a \cos n\theta$ is an n-leaf rose if n is odd, or a *2n*-leaf rose if n is even.

Example 3

Sketch the graph of $r = 2 + 2 \sin \theta.$

Solution: We will begin with a graph of $r = 2 + 2 \sin \theta$ in rectangular coordinates, as shown in Figure 7.14.

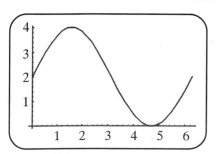

Figure 7.14.

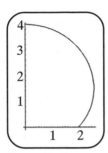

Figure 7.15.

Notice that in the rectangular graph of $r = 2 + 2$ sin θ, r is always positive. When $\theta = 0$, $r = 2$, and as θ increases from 0 to $\frac{\pi}{2}$, sweeping out Quadrant I, r increases from 2 to 4. The graph of this region is shown in Figure 7.15.

As θ increases from $\frac{\pi}{2}$ to $\frac{3\pi}{2}$, and sweeps out Quadrant II and Quadrant III, r decreases from 4 to 0. Figure 7.16 shows the graph of this region.

As θ increases from $\frac{3\pi}{2}$ to 2π, r increases from 0 to 2. The graph of this region is shown in Figure 7.17.

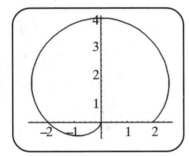

Figure 7.16.

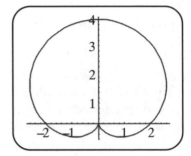

Figure 7.17.

The heart-shaped curve in Example 3 is called a cardioid. In general, any equation of the form $r = a(1 \pm \sin \theta)$ and $r = a(1 \pm \cos \theta)$ is a cardioid.

Example 4

Sketch the graph of $r = 1 + 2 \sin \theta$.

Solution: Begin by graphing $r = 1 + 2 \sin \theta$ in rectangular coordinates, as shown in Figure 7.18.

Notice that this graph crosses the θ-axis at $\theta = \frac{7\pi}{6}$ and $\theta = \frac{11\pi}{6}$. To see this, set $r = 0$ and solve for θ:

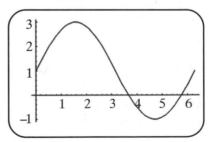

Figure 7.18.

$1 + 2\sin\theta = 0$

$2\sin\theta = -1$

$\sin\theta = -\frac{1}{2}$

$\theta = \frac{7\pi}{6}$ or $\theta = \frac{11\pi}{6}$

These angles are worth noting, because the sign of r changes at these angles. We will need to pay attention to this when we sketch the graph as θ ranges from $\frac{7\pi}{6}$ to $\frac{11\pi}{6}$.

When $\theta = 0$, $r = 1$, and as θ increases from 0 to $\frac{\pi}{2}$, sweeping out Quadrant I, r increases from 2 to 3. The graph of this region is shown in Figure 7.19.

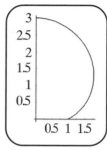

Figure 7.19.

As θ increases from $\frac{\pi}{2}$ to π, and sweeps out Quadrant II, r decreases from 3 to 1. The graph of this region is shown in Figure 7.20.

As θ increases from π to $\frac{7\pi}{6}$, r decreases from 1 to 0. The graph of this region is shown in Figure 7.21.

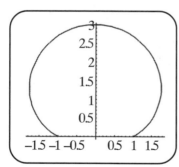

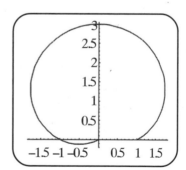

Figure 7.20.   *Figure 7.21.*

As θ increases from $\frac{7\pi}{6}$ to $\frac{3\pi}{2}$, and θ sweeps out the rest of Quadrant III, r decreases from 0 to -1. The effect of the negative value of r is that the graph moves back to Quadrant I, as shown in Figure 7.22.

As θ increases from $\frac{3\pi}{2}$ to $\frac{11\pi}{6}$, and θ sweeps out part of Quadrant IV, r increases from -1 to 0. The effect of the negative value of r is that the graph moves back to Quadrant II, as shown in Figure 7.23.

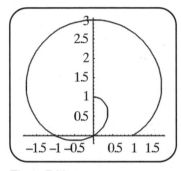

Figure 7.22.

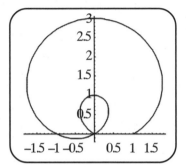

Figure 7.23.

As θ increases from $\frac{11\pi}{6}$ to 2π, and θ sweeps out the rest of Quadrant IV, r increases from 0 to 1.

The complete graph of $r = 1 + 2$ sin θ is shown in Figure 7.24.

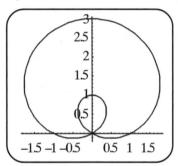

Figure 7.24.

To graph polar equations, I recommend starting with the graphs of the function in rectangular coordinates, using an $r - \theta$ set of axes, and then interpreting that information to draw the graph in the polar grid. It is a skill that involves a change in perspective, and it will take some practice to develop.

Lesson 7-2: Review

1. Sketch the graph of $r = 3\cos \theta$.

2. Sketch the graph of $r = \cos 2\theta$.

3. Sketch the graph of $r = 2 + 2\cos \theta$.

4. Sketch the graph of $r = 1 + 2\cos \theta$.

Lesson 7-3: The Polar Form of Complex Numbers

Trigonometry can be used to solve problems in geometry that involve triangles. It can be used to model the periodic properties of sunlight and sound waves. It would appear that its applications are limited to geometry and repetitive behavior, but there are other ways that trigonometry can be used. As we will learn in this lesson, trigonometry can be used to simplify algebraic calculations that involve complex numbers.

There are no real solutions to the equation $x^2 + 1 = 0$. If we try to solve this equation for x, we end up with the equation $x^2 = -1$, and there is no real number that, when squared, becomes negative: the square of any real number is positive. This does not mean that it is impossible to solve the equation $x^2 + 1 = 0$. It means that any solution to that equation cannot be a real number.

Mathematicians invented an expanded number system, called the **complex number system**, which makes it possible to find solutions to every quadratic equation. First, define the number i to be a solution to the equation $x^2 + 1 = 0$. If we substitute i into this equation, we see that $(i)^2 + 1 = 0$, or $i^2 = -1$. The powers of i form a pattern shown in the adjacent table.

From this, we see that odd powers of i are $\pm i$, and even powers of i are ± 1. As a result, the highest power of i in any expression should be 1.

A **complex number** is an expression of the form $a + bi$, where a and b are real numbers and i satisfies the equation $i^2 = -1$. In the expression

Power of i	Expression
1	i
2	$i^2 = -1$
3	$i^3 = i^2 \cdot i = (-1) \cdot i = -i$
4	$i^4 = i^2 \cdot i^2 = (-1) \cdot (-1) = 1$
5	$i^5 = i^4 \cdot i = 1 \cdot i = i$
6	$i^6 = i^4 \cdot i^2 = 1 \cdot (-1) = -1$
7	$i^7 = i^4 \cdot i^3 = 1 \cdot (-i) = -i$
8	$i^8 = i^4 \cdot i^4 = 1 \cdot 1 = 1$

$a + bi$, a is called the **real part**, and b is called the **imaginary part**. It is important to remember that the *imaginary* part of a complex number is actually a *real* number: it is the *coefficient* of i. Two complex numbers are equal if their real parts are equal and their imaginary parts are equal. Every real number can be written as a complex number: the real number a can be written as $a + 0i$.

The table that follows on page 158 gives examples of complex numbers and specifies their real and imaginary parts.

Complex Number	Real Part	Imaginary Part
$3 + 2i$	3	2
$4 - 3i$	4	-3
$2i$	0	2
-5	-5	0

The number $2i$ is called a **pure imaginary number**. A pure imaginary number is a complex number whose real part is 0. Complex numbers are usually denoted by z, and it is common to write a complex number as $z = x + iy$.

Complex numbers can be combined algebraically. They can be added, subtracted, multiplied, and divided. To add or subtract complex numbers, combine the real parts and the imaginary parts. Multiplying two complex numbers together involves the same process as multiplying two binomials. There is one extra step, however. You will make use of the fact that $i^2 = -1$.

Example 1

Simplify the following expressions:

a. $(2 + 3i) + (9 - 2i)$

b. $(2 + 3i) - (9 - 2i)$

c. $(2 + 3i)(9 - 2i)$

d. $(2 + 3i)(2 - 3i)$

Solution: Combine the real and imaginary terms as appropriate. Use the distributive law for subtraction, and use the FOIL method for multiplying two binomials.

a. $(2 + 3i) + (9 - 2i) = 11 + i$

b. $(2 + 3i) - (9 - 2i) = -7 + 5i$

c. $(2 + 3i)(9 - 2i) = 18 - 6i^2 - 4i + 27i = 18 + 6 - 4i + 27i = 24 + 23i$

d. $(2 + 3i)(2 - 3i) = 4 - 9i^2 - 6i + 6i = 4 + 9 = 13$

The last part of Example 1 is very interesting. The product of these two complex numbers is a real number. The two complex numbers that are being multiplied together are very similar. The only difference between them is that the signs of their imaginary parts are opposites. It can be useful to know how to transform a complex number into a real number. The complex numbers $(2 + 3i)$ and $(2 - 3i)$ are called *complex conjugates* of each other. In general, the **complex conjugate** of $a + bi$ is $a - bi$. Multiplication by a conjugate is part of the process of dividing one complex number by another complex number. This technique should look familiar to you. The process of rationalizing the denominator of a fraction whose denominator involves a radical involves multiplying by the conjugate of the expression in the denominator.

Example 2

Simplify the expression $\dfrac{2+3i}{1-2i}$.

Solution: Simplifying this expression involves multiplying both the numerator and the denominator by the complex conjugate of the denominator and simplifying:

$$\frac{2+3i}{1-2i}$$

Multiply both the numerator and the denominator by $(1 + 2i)$

$$\frac{(2+3i)}{(1-2i)} \cdot \frac{(1+2i)}{(1+2i)}$$

Expand both products
$$\frac{2+6i^2+4i+3i}{1-4i^2-4i+4i}$$

Use the fact that $i^2 = -1$
$$\frac{2-6+4i+3i}{1+4-4i+4i}$$

$$\frac{-4+7i}{5} = -\frac{4}{5}+\frac{7}{5}i$$

It would seem that working with complex numbers has very little to do with trigonometry. As it turns out, there are other algebraic operations, such as squaring, cubing, and taking the square root of a complex number, that we can do more easily with the help of trigonometry. We will need to look at complex numbers through trigonometric eyes.

Complex numbers have two components: a real part and an imaginary part. We can think of a complex number as an ordered pair (a, b), where a represents the real part and b represents the imaginary part of the complex number. We can then consider the collection of complex numbers as a two-dimensional plane, similar to the Cartesian coordinate plane. The set of complex numbers lie in the **complex plane**, as it is called. In the complex plane, the horizontal axis is called the **real axis**, and the vertical axis is called the **imaginary axis**.

I mentioned earlier that complex numbers are often written as $z = x + iy$. The Cartesian coordinate system and the complex plane are so closely related that this notation is natural. The real axis in the complex plane corresponds to the x-axis in the Cartesian coordinate plane, and the imaginary axis in the complex plane corresponds to the y-axis in the Cartesian coordinate plane. We can graph complex numbers in the complex plane the same way we plotted a point in the Cartesian coordinate plane.

Example 3

Graph the complex numbers $z_1 = 3 + i$ and $z_2 = 2 - i$ in the complex plane.

Solution: The graph of the complex numbers $3 + i$ and $2 - i$ are the points (3, 1) and (2, –1), as shown in Figure 7.25.

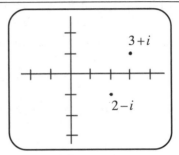

Figure 7.25.

Now that we can visualize a complex number as an ordered pair in the complex plane, we can see how trigonometry can be used. A complex number in the complex plane can be used to generate a right triangle, where the length of one leg is the absolute value of the real part of the complex number and the length of the other leg is the absolute value of the imaginary part of the complex number. We must take the absolute value of each component because distance is always a positive number. The distance between a complex number and the origin corresponds to the length of the hypotenuse of this right triangle, and the hypotenuse of this triangle and the positive real axis will create an angle θ. All of the right triangle trigonometry that we developed in Chapter 2 can be interpreted in terms of complex numbers.

The distance between a complex number and the origin is called the **modulus** of the complex number. If $z = x + iy$ is a complex number, then the modulus of z, denoted $|z|$, is defined using the Pythagorean Theorem:

$$|z| = \sqrt{x^2 + y^2}$$

The modulus of a complex number $z = x + iy$ is the same as the radius of the polar coordinate representation of the point (x, y).

Example 4

Find the modulus of the complex numbers $z_1 = 3 + i$ and $z_2 = 2 - i$.

Solution: Use the Pythagorean Theorem to find the distance between each point and the origin. This is equivalent to finding the r-coordinate of the polar coordinate representation of the points $(3, 1)$ and $(2, -1)$:

$$|z_1| = \sqrt{3^2 + 1^2} = \sqrt{9 + 1} = \sqrt{10}$$

$$|z_2| = \sqrt{2^2 + (-1)^2} = \sqrt{4 + 1} = \sqrt{5}$$

It should not surprise you that complex numbers have a polar representation. The polar representation of a complex number is consistent with the polar representation for a point in the Cartesian coordinate system. If $z = x + iy$ is a complex number in the complex plane, then the line segment connecting the origin and this point forms an angle θ relative to the positive real axis. If r is the distance between the origin and the point, then from the definitions of the sine and cosine ratios, we see that:

$x = r \cos \theta$ and $y = r \sin \theta$

and:

$z = r \cos \theta + ir \sin \theta = r(\cos \theta + i \sin \theta)$

In other words, the polar representation of the complex number $z = x + iy$ is $z = r(\cos \theta + i \sin \theta)$, where r is the modulus of z ($r = |z|$) and $\tan \theta = \frac{y}{x}$. The angle θ is called the **argument** of z. Although the argument of z will not be unique, any two arguments of the same complex number must differ by an integer multiple of 2π.

Example 5

Write the following complex number in trigonometric form:

a. $2 - 2i$ b. $-3 + 4i$

Solution: Find the modulus, $r = |z|$, and the argument, $\tan \theta = \frac{y}{x}$, of each complex number and then write the number in the form $z = r(\cos \theta + i \sin \theta)$. Remember that the argument is not unique.

a. $2 - 2i$: The modulus is $|z| = \sqrt{2^2 + (-2)^2} = 2\sqrt{2}$ and the argument

is $\theta = \tan^{-1}\left(-\frac{2}{2}\right) = \tan^{-1}(-1)$. The complex number $2 - 2i$ is

located in Quadrant IV, so $\theta = \tan^{-1}(-1) = -\frac{\pi}{4}$. Therefore,

$2 - 2i = 2\sqrt{2}\left(\cos\left(-\frac{\pi}{4}\right) + i\sin\left(-\frac{\pi}{4}\right)\right)$. We can clean this equation

up by using the symmetry of the sine and cosine functions:

$2 - 2i = 2\sqrt{2}\left(\cos\frac{\pi}{4} - i\sin\frac{\pi}{4}\right)$.

b. $-3 + 4i$: The modulus is $|z| = \sqrt{(-3)^2 + 4^2} = 5$ and the argument

is $\theta = \tan^{-1}\left(-\frac{4}{3}\right)$. This is not one of our special angles, so we

cannot find the exact value of this angle. Keep in mind that the complex number $-3 + 4i$ is located in Quadrant II, which may be a useful piece of information. Putting these pieces together,

we have: $-3 + 4i = 5\left(\cos\left[\tan^{-1}\left(-\frac{4}{3}\right)\right] + i\sin\left[\tan^{-1}\left(-\frac{4}{3}\right)\right]\right)$.

I mentioned earlier that multiplying, dividing, squaring, cubing, and taking the square root of complex numbers can be done more easily by looking at complex numbers through trigonometric eyes. The key to this interpretation is the following equation:

$e^{i\theta} = \cos\theta + i\sin\theta.$

The expression $e^{i\theta}$ may look strange, yet familiar, to you. It should remind you of exponential functions of the form $f(x) = e^x$. We can use this compact representation for $\cos\theta + i\sin\theta$ and apply the properties of exponents:

$e^{m+n} = e^m e^n$

$e^{m-n} = \dfrac{e^m}{e^n}$

$(e^m)^n = e^{mn}$

The advantage to working with expressions of the form $e^{i\theta}$ is that the rules for exponents can be used to simplify calculations. In general, if $z = x + iy$, and the modulus of z is r and the argument of z is θ, then

$z = re^{i\theta}$. We can refer to the expression $re^{i\theta}$ as the **exponential form** of a complex number. From this, we see that $z^n = (re^{i\theta})^n = r^n e^{in\theta}$. Equivalently, $z^n = r^n(\cos n\theta + i \sin n\theta)$.

> The formula $z^n = r^n(\cos n\theta + i \sin n\theta)$ is known as
> **DeMoivre's Theorem.**

Example 6

Expand the expression $(2 - 2i)^4$.

Solution: It would be tedious to expand the expression $(2 - 2i)^4$ by evaluating $(2 - 2i)(2 - 2i)(2 - 2i)(2 - 2i)$. In Example 5, we saw that the modulus of $(2 - 2i)$ is $\sqrt{8}$, and the argument of z is $-\frac{\pi}{4}$. Using the formula $z^n = r^n(\cos n\theta + i \sin n\theta)$, we see that:

$$(2-2i)^4 = \left(\sqrt{8}\right)^4 \left(\cos\left[4 \cdot \left(-\tfrac{\pi}{4}\right)\right] + i\sin\left[4 \cdot \left(-\tfrac{\pi}{4}\right)\right]\right)$$

We can clean this equation up:

$(2 - 2i)^4 = 64(\cos[-\pi] + i \sin[-\pi])$

$(2 - 2i)^4 = 64(-1 + i \cdot 0)$

$(2 - 2i)^4 = -64$

We could have also solved this problem by simplifying the equation $z^n = r^n e^{in\theta}$ directly, and then putting it in trigonometric form:

$$(2-2i)^4 = \left(\sqrt{8}\right)^4 \left(e^{-\frac{i\pi}{4}}\right)^4$$

$(2 - 2i)^4 = 64e^{-i\pi}$

$(2 - 2i)^4 = 64(\cos[-\pi] + i \sin[-\pi])$

$(2 - 2i)^4 = 64(-1 + i \cdot 0)$

$(2 - 2i)^4 = -64$

It doesn't matter which form you start with. The end result is the same, and the process is considerably simpler by using the exponential form of a complex number.

The exponential form of a complex number can also be used to derive formulas for multiplying and dividing complex numbers.

If $z_1 = r_1(\cos\theta_1 + i\sin\theta_1)$ and $z_2 = r_2(\cos\theta_2 + i\sin\theta_2)$, then:

$$z_1 \cdot z_2 = r_1 r_2 \left(\cos(\theta_1 + \theta_2) + i\sin(\theta_1 + \theta_2)\right)$$

$$\frac{z_1}{z_2} = \frac{r_1}{r_2}\left(\cos(\theta_1 - \theta_2) + i\sin(\theta_1 - \theta_2)\right)$$

These formulas should come as no surprise, and may be difficult to remember unless you understand that they come from the exponential form of a complex number. If $z_1 = r_1 e^{i\theta_1}$ and $z_2 = r_2 e^{i\theta_2}$, then using the properties of exponents, we have:

$$z_1 \cdot z_2 = r_1 e^{i\theta_1} \cdot r_2 e^{i\theta_2} = r_1 \cdot r_2 e^{i(\theta_1 + \theta_2)}$$

$$\frac{z_1}{z_2} = \frac{r_1 e^{i\theta_1}}{r_2 e^{i\theta_2}} = \frac{r_1}{r_2} e^{i(\theta_1 - \theta_2)}$$

These equations can be easily transformed into their trigonometric equivalents:

$$z_1 \cdot z_2 = r_1 r_2 \left(\cos(\theta_1 + \theta_2) + i\sin(\theta_1 + \theta_2)\right)$$

$$\frac{z_1}{z_2} = \frac{r_1}{r_2}\left(\cos(\theta_1 - \theta_2) + i\sin(\theta_1 - \theta_2)\right)$$

Example 7

If $z_1 = 2\left(\cos\frac{\pi}{3} + i\sin\frac{\pi}{3}\right)$ and $z_2 = 3\left(\cos\frac{\pi}{6} + i\sin\frac{\pi}{6}\right)$, find the following:

a. $z_1 \cdot z_2$ b. $\dfrac{z_1}{z_2}$ c. $(z_1)^9$

Solution: Use the formulas we just derived:

a. $z_1 \cdot z_2 = 2 \cdot 3\left(\cos\left(\frac{\pi}{3} + \frac{\pi}{6}\right) + i\sin\left(\frac{\pi}{3} + \frac{\pi}{6}\right)\right) = 6\left(\cos\frac{\pi}{2} + i\sin\frac{\pi}{2}\right) = 6i$

b. $\dfrac{z_1}{z_2} = \frac{2}{3}\left(\cos\left(\frac{\pi}{3} - \frac{\pi}{6}\right) + i\sin\left(\frac{\pi}{3} - \frac{\pi}{6}\right)\right) = 6\left(\cos\frac{\pi}{6} + i\sin\frac{\pi}{6}\right)$

c. $(z_1)^9 = 2^9\left(\cos\frac{9\pi}{3} + i\sin\frac{9\pi}{3}\right) = 512(\cos 3\pi + i\sin 3\pi) = -512$

The equation $z^n = r^n e^{in\theta}$ can also be used to find various roots of a complex number. There is a subtle difference between finding the roots of a complex number and finding the roots of a real number. A real number has two square roots: a positive root and a negative root. For example, -2 and 2 are both square roots of 4, because $2^2 = 4$ and $(-2)^2 = 4$. Whereas a *real* number only has one cube root, a *complex* number will actually have three distinct cube roots. In general, a complex number will have n distinct nth roots, and we can find them by using the following formula.

If $z = r(\cos\theta + i\sin\theta)$ and n is a positive integer, then the n distinct nth roots of z can be found using the equation:

$$w_k = r^{1/n}\left[\cos\left(\tfrac{\theta+2k\pi}{n}\right)+i\sin\left(\tfrac{\theta+2k\pi}{n}\right)\right], \text{ where } k = 0, 1, 2, 1/4, n - 1.$$

In other words, the modulus of each nth root is $r^{1/n}$, and the argument of the first root is $\frac{\theta}{n}$. To find the next root, add $\frac{2\pi}{n}$ to the argument of the previous root. If you graph the n distinct roots of a complex number, the roots will be evenly spaced around the circle of radius $r^{1/n}$.

Example 8

Find the five fifth roots of -32.

Solution: First, write the polar form of the complex number -32. The modulus is 32 and the argument is π. $z = 32(\cos\pi + i\sin\pi)$. Find the first fifth root:

$$w_0 = 2\left[\cos\left(\tfrac{\pi}{5}\right)+i\sin\left(\tfrac{\pi}{5}\right)\right]$$

The argument of the first fifth root is $\frac{\pi}{5}$. To find the second fifth root, add $\frac{2\pi}{5}$ to $\frac{\pi}{5}$:

$$w_1 = 2\left[\cos\left(\tfrac{3\pi}{5}\right)+i\sin\left(\tfrac{3\pi}{5}\right)\right].$$

Continue in this manner and go around the circle:

$$w_2 = 2\left[\cos(\pi)+i\sin(\pi)\right]$$

$$w_3 = 2\left[\cos\left(\tfrac{7\pi}{5}\right) + i\sin\left(\tfrac{7\pi}{5}\right)\right]$$

$$w_4 = 2\left[\cos\left(\tfrac{9\pi}{5}\right) + i\sin\left(\tfrac{9\pi}{5}\right)\right]$$

At this point, if we added another $\tfrac{2\pi}{5}$ to the argument of the current root ($\tfrac{9\pi}{5}$), we would have $\tfrac{11\pi}{5}$, which is just $\tfrac{\pi}{5} + 2\pi$, so we are back to the first fifth root that we already found. The five fifth roots of -32 are shown in Figure 7.26.

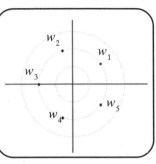

Figure 7.26.

We can use the formula for finding the nth roots of a complex number to solve algebraic equations. For example, the solutions to the equation $z^5 + 32 = 0$ are the five fifth roots of -32, which we found in the previous example.

Lesson 7-3 Review

1. Simplify the following expressions:

   a. $(4 - 6i) - (5 - 2i)$ 　　　　c. $(1 + 5i)(2 - i)$

   b. $(3 + 2i)(3 - 2i)$ 　　　　d. $\dfrac{(1+5i)}{(2-i)}$

2. Write the complex number $\sqrt{3} - i$ in trigonometric form.

3. If $z_1 = 3\left(\cos\tfrac{\pi}{4} + i\sin\tfrac{\pi}{4}\right)$ and $z_2 = 4\left(\cos\tfrac{\pi}{3} + i\sin\tfrac{\pi}{3}\right)$, find the following:

   a. $z_1 \cdot z_2$ 　　　　b. $\dfrac{z_1}{z_2}$ 　　　　c. $(z_1)^5$

4. Find the three cube roots of i.

Answer Key

Lesson 7-1 Review

1. $\left(-1,\sqrt{3}\right)$

2. $\left(3\sqrt{2},\frac{3\pi}{4}\right)$

3. a. $x^2 + y^2 = 9$ b. $x = \frac{\sqrt{3}}{3}|y|$

4. $\tan\theta = 1$

Lesson 7-2 Review

1. The graph of $r = 3\cos\theta$ is shown in Figure 7.27.

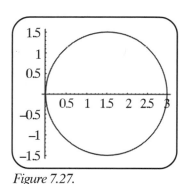

Figure 7.27.

2. The graph of $r = \cos 2\theta$ is shown in Figure 7.28.

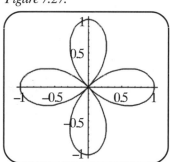

Figure 7.28.

3. The graph of $r = 2 + 2\cos\theta$ is shown in Figure 7.29.

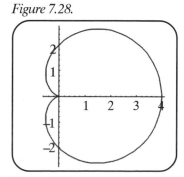

Figure 7.29.

4. The graph of $r = 1 + 2\cos\theta$ is shown in Figure 7.30.

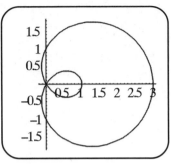

Figure 7.30

Lesson 7-3 Review

1. a. $-1 - 4i$

   b. 13

   c. $7 + 9i$

   d. $-\frac{3}{5} + \frac{11}{5}i$

2. $2\left(\cos\frac{\pi}{6} - i\sin\frac{\pi}{6}\right)$

3. a. $z_1 \cdot z_2 = 12\left(\cos\frac{7\pi}{12} + i\sin\frac{7\pi}{12}\right)$

   b. $\dfrac{z_1}{z_2} = \dfrac{3}{4}\left(\cos\frac{\pi}{12} - i\sin\frac{\pi}{12}\right)$

   c. $\left(z_1\right)^5 = 243\left(\cos\frac{5\pi}{4} + i\sin\frac{5\pi}{4}\right)$

4. $i = e^{\frac{\pi i}{2}}$, so $w_k = \cos\frac{\frac{\pi}{2} + 2k\pi}{3} + i\sin\frac{\frac{\pi}{2} + 2k\pi}{3}$:

   $w_1 = \cos\frac{\pi}{6} + i\sin\frac{\pi}{6}$

   $w_2 = \cos\frac{5\pi}{6} + i\sin\frac{5\pi}{6}$

   $w_3 = \cos\frac{3\pi}{2} + i\sin\frac{3\pi}{2}$

Vectors

Many measurable aspects of matter, such as temperature, mass, and energy, only depend their size, or magnitude. These quantities are represented by a single number, and are called **scalars**. In order to describe the *movement* of an object in the coordinate plane or in three-dimensional space, *two* numbers are required: the length of the distance that is traveled as well as the direction of the movement. A quantity that is described using both a magnitude and a direction are called **directed quantities**. The velocity and acceleration of an object are examples of directed quantities. Directed quantities can be represented mathematically by using vectors. In this chapter, we will develop rules for describing and manipulating vectors. We will also look at vectors from a trigonometric perspective.

Lesson 8-1: Vectors

A **vector** is a quantity that has both magnitude and direction. A vector in the plane is represented as a line segment with an assigned direction, or, more familiarly, as an arrow. The length of the arrow represents the magnitude of the vector, and the direction of the arrow specifies the direction of the vector. The notation for a vector is the same as the notation used in geometry to represent a ray. Figure 8.1 shows the vector $\overrightarrow{AB}$. The point A is called the **initial point**, and B is the **terminal point** of the vector. The length of the line segment AB is called the **magnitude**, or **length**, or

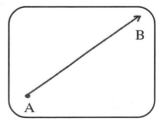

Figure 8.1.

norm of the vector, and is denoted by $\left|\overrightarrow{AB}\right|$. Vectors are usually written in bold, to distinguish between vectors and scalars.

Two vectors are **equivalent** if they have the same magnitude and point in the same direction. Because a vector is determined solely by its length and direction, equivalent vectors are considered to be equal even though their positions may be different. Translating a vector will not change the vector, but rotating, stretching, or contract-

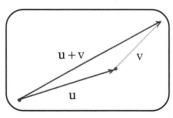

Figure 8.2.

ing the vector will. Two vectors can be combined through addition and subtraction. If **u** and **v** are two vectors, we can find their sum geometrically by sketching vectors equal to **u** and **v** with the initial point of one vector coinciding with the terminal point of the other vector, as shown in Figure 8.2.

Another way to visualize vector addition is to sketch both vectors so that their initial points coincide. Create a parallelogram using the two vectors. The vector **u** + **v**, or the sum of the two vectors, is the vector that coincides with the diagonal of the parallelogram that includes the common initial point of **u** and **v**. This method for vector addition is illustrated in Figure 8.3.

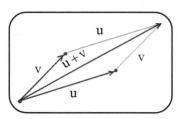

Figure 8.3.

Subtracting one vector from another vector can also be done by constructing a parallelogram. The vector **u** − **v** is the diagonal of the parallelogram whose initial point is the head of **v** and whose terminal point is the head of **u**, as shown in Figure 8.4.

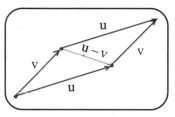

Figure 8.4.

Given a vector **v**, the vector −**v** is a vector that has the same magnitude as **u** and that points in the opposite direction. The vector −**v** is called the additive inverse of **v**. The sum of the vectors **v** and −**v** is the **zero** vector, or the vector with zero length.

The vector **v** + **v** can be written as the vector 2**v**. It is a vector that points in the same direction as **v** but whose magnitude is twice the magnitude of **v**. The generalization of this process is called scalar multiplication. The vector a**v** has magnitude $|a|\,|\mathbf{v}|$ and points in the same direction as **v**.

Multiplying by a positive scalar will stretch the vector, if the scalar is greater than 1, or contract the vector, if the scalar is less than 1. If the scalar is negative, then the direction of $a\mathbf{v}$ is the opposite of the direction of $\mathbf{v}$.

If we draw a vector in the coordinate plane, then we can work with the coordinates of the vector. A vector can be specified by giving the location of its initial and terminal points, as shown in Figure 8.5. The direction of the vector can be described using the slope of the line segment. If the coordinates of the terminal point are $(x_{\text{terminal}}, y_{\text{terminal}})$, and the coordinates of the initial point are $(x_{\text{initial}}, y_{\text{initial}})$, then the slope of the line segment can be found using the formula:

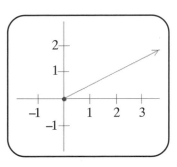

Figure 8.5.

$$\text{slope} = \frac{y_{\text{terminal}} - y_{\text{initial}}}{x_{\text{terminal}} - x_{\text{initial}}}$$

If we let $a = x_{\text{terminal}} - x_{\text{initial}}$, and $b = y_{\text{terminal}} - y_{\text{initial}}$, then we can represent the vector $\mathbf{u}$ as an ordered pair of real numbers:

$$\mathbf{v} = \langle a, b \rangle = \langle x_{\text{terminal}} - x_{\text{initial}}, y_{\text{terminal}} - y_{\text{initial}} \rangle$$

The notation $\mathbf{v} = \langle a, b \rangle$ is called the component form of a vector. We will use $\langle\ ,\ \rangle$ to represent a vector, and $(\ ,\ )$ for the ordered pair representation of a point. Two vectors are parallel if one vector is a scalar multiple of the other. In other words, $\mathbf{u}$ and $\mathbf{v}$ are parallel if there is a scalar c so that $\mathbf{u} = c\mathbf{v}$.

Drawing a vector in the coordinate plane should remind you of right triangle trigonometry. For vectors that are not parallel to the coordinate axes, we can construct a right triangle where the vector forms the hypotenuse of the right triangle. Using the component form of a vector, we can derive formulas for the magnitude and direction of a vector. For the vector $\mathbf{v} = \langle a, b \rangle$, the magnitude of the vector can be found using the equation

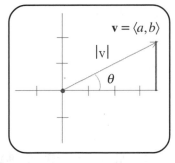

Figure 8.6.

$|v| = \sqrt{a^2 + b^2}$. The direction of the vector can be specified by the angle shown in Figure 8.6.

The direction of a vector is the smallest positive angle in standard position that is formed by the positive x-axis and **v**.

The magnitude of a vector satisfies the following four important properties:

The magnitude of a vector must be non-negative.	$\lvert \mathbf{v} \rvert \geq 0$
The zero vector is the only vector whose magnitude is 0.	$\lvert \mathbf{v} \rvert = 0$ if and only if $\mathbf{v} = 0$
The magnitude of a vector is equal to the magnitude of its additive inverse.	$\lvert -\mathbf{v} \rvert = \lvert \mathbf{v} \rvert$
The magnitude of the product of a scalar and a vector is the product of the magnitudes of the scalar and the vector.	$\lvert c\mathbf{v} \rvert = \lvert c \rvert \lvert \mathbf{v} \rvert$

This should remind you of how we developed complex numbers in the last chapter. Once we have a coordinate representation for an object, then all of the trigonometric results will apply to this new situation. A vector is in **standard position** if its initial point coincides with the origin of the coordinate plane.

We can use the component representation to add and subtract vectors easily. Instead of having to draw the two vectors and align them head to tail, or construct a parallelogram, we can just add (or subtract) the *components* of the two vectors. Scalar multiplication involves multiplying each component of the vector by the particular scalar. We can summarize these properties as follows. If $\mathbf{v}_1 = \langle a_1, b_1 \rangle$ and $\mathbf{v}_2 = \langle a_2, b_2 \rangle$ are vectors, and c is a scalar (or more commonly referred to as a real number), then:

$$\mathbf{v}_1 + \mathbf{v}_2 = \langle a_1 + a_2, b_1 + b_2 \rangle$$
$$\mathbf{v}_1 - \mathbf{v}_2 = \langle a_1 - a_2, b_1 - b_2 \rangle$$
$$c\mathbf{v}_1 = \langle c \cdot a_1, c \cdot b_1 \rangle$$

Example 1

If $\mathbf{u} = \langle 3, -1 \rangle$ and $\mathbf{v} = \langle 5, -2 \rangle$, find the following:

a. $\mathbf{u} + \mathbf{v}$ b. $\mathbf{u} - \mathbf{v}$ c. $3\mathbf{u}$

Solution: Use the components of each vector:

a. $\mathbf{u} + \mathbf{v} = \langle 3, -1 \rangle + \langle 5, -2 \rangle = \langle 8, -3 \rangle$

b. $\mathbf{u} - \mathbf{v} = \langle 3, -1 \rangle - \langle 5, -2 \rangle = \langle -2, 1 \rangle$

c. $3\mathbf{u} = 3\langle 3, -1 \rangle = \langle 9, -3 \rangle$

Whenever a new mathematical object is presented, it is a good idea to spend some time discussing some of its properties and establishing some rules for how to work with it. Vectors have some important properties that are specified in the table that follows.

Vector addition commutes.	$\mathbf{u} + \mathbf{v} = \mathbf{v} + \mathbf{u}$
Vector addition is associative.	$\mathbf{u} + (\mathbf{v} + \mathbf{w}) = (\mathbf{u} + \mathbf{v}) + \mathbf{w}$
There is a zero vector, which is denoted by **0.**	$\mathbf{v} + \mathbf{0} = \mathbf{v}$
Every vector **v** has an additive inverse, denoted by **−v.**	$\mathbf{v} + (-\mathbf{v}) = \mathbf{0}$
Multiplication of a vector by a scalar distributes over vector addition.	$c(\mathbf{u} + \mathbf{v}) = c\mathbf{u} + c\mathbf{v}$
Multiplication of a scalar and a vector distributes over scalar addition.	$(c + d)\,\mathbf{v} = c\mathbf{v} + d\mathbf{v}$
Scalar multiplication is commutative and associative.	$(cd)\mathbf{v} = c(d\mathbf{v}) = d(c\mathbf{v})$
There is an identity for scalar multiplication.	$1 \cdot \mathbf{v} = \mathbf{v}$
Multiplication by the scalar 0 gives the zero vector.	$0 \cdot \mathbf{v} = \mathbf{0}$
Scalar multiplication of the zero vector is the zero vector.	$c \cdot \mathbf{0} = \mathbf{0}$

A **unit** vector is a vector whose length is 1. Given a nonzero vector **v**, it is possible to find a unit vector that has the same direction as **v**. The vector given by $\mathbf{u} = \dfrac{\mathbf{v}}{|\mathbf{v}|}$ is a unit vector having the same direction as **v**.

Two very important unit vectors are the vectors $\langle 1, 0 \rangle$ and $\langle 0, 1 \rangle$. The vector $\langle 1, 0 \rangle$ is parallel to the x-axis, and the vector $\langle 0, 1 \rangle$ is parallel to the y-axis. These vectors are used extensively in physics, and are given special names: $\mathbf{i} = \langle 1, 0 \rangle$ and $\mathbf{j} = \langle 0, 1 \rangle$. Any vector in the plane can be easily written in terms of these vectors. The vector $\langle a, b \rangle$ can be written as $a\mathbf{i} + b\mathbf{j}$. The scalar a is called the **horizontal component** of the vector, and b is called the **vertical component** of the vector. For example, the vector $\langle 5, -3 \rangle$ can be written as $5\mathbf{i} - 3\mathbf{j}$; the horizontal component of $\langle 5, -3 \rangle$ is 5, and its vertical component is -3. Do not confuse the vector $\mathbf{i}$ and the complex number i. They are very different mathematical objects.

We are ready to make the connection between vector representations and trigonometry. Suppose $\mathbf{v}$ is a vector with magnitude $|\mathbf{v}|$ and direction θ. Then $\mathbf{v} = \langle a, b \rangle = a\mathbf{i} + b\mathbf{j}$, where $a = |\mathbf{v}|\cos\theta$ and $b = |\mathbf{v}|\sin\theta$. In other words, we can write a vector $\mathbf{v}$ as $\mathbf{v} = |\mathbf{v}|\cos\theta\,\mathbf{i} + |\mathbf{v}|\sin\theta\,\mathbf{j}$. This equation enables us to find the components of a vector if we know its magnitude and its direction.

Example 2

Find the horizontal and vertical components of a vector whose magnitude is 4 and direction is $\frac{3\pi}{4}$.

Solution: The horizontal component of the vector is

$|\mathbf{v}|\cos\theta = 4\cos\frac{3\pi}{4} = -2\sqrt{2}$, and the vertical component is

$|\mathbf{v}|\sin\theta = |\mathbf{v}|\sin\frac{3\pi}{4} = 2\sqrt{2}$.

Example 3

A ball is thrown with an initial speed of 35 mph in a direction that makes an angle of 30° with the positive x-axis. Find the initial speed in the horizontal direction, and the initial speed in the vertical direction.

Solution: The initial velocity in the horizontal direction is the horizontal component of the velocity vector:

$$|\mathbf{v}|\cos\theta = 35\cos 30° = \frac{35\sqrt{3}}{2} \approx 30.3 \text{ mph}$$

The initial velocity in the vertical direction is the vertical component of the velocity vector:

$$|v|\sin\theta = 35\sin 30° = \frac{35}{2} = 17.5 \text{ mph}$$

Example 4

A boat is launched from one shore of a straight river. Its destination is a point directly on the opposite shore. If the speed of the boat relative to the water is 10 mph, and the river is flowing east at a rate of 5 mph, in what direction should the boat travel in order to arrive at the desired landing point?

Solution: A graph of this situation, as shown in Figure 8.7, may be helpful.

We are given the magnitude of the velocity of the boat, and the velocity vector of the river. We need to find the direction of the velocity of the boat. We need an orientation for our boat and the river. We can use the vector **i** to represent a velocity whose direction is east, and **j** to represent a velocity whose direction is north. Let **u** represent the velocity of the river, and **v** represent the velocity of the boat. Then **u** = 5**i**. The magnitude of the velocity of the boat is 10, and the argument of the velocity of the boat is unknown. Let θ represent the argument of the velocity of the boat. Then

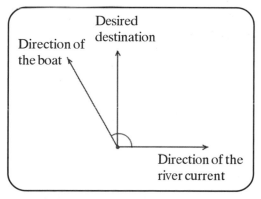

Figure 8.7.

$$\mathbf{v} = |v|\cos\theta\,\mathbf{i} + |v|\sin\theta\,\mathbf{j} = 10\cos\theta\,\mathbf{i} + 10\sin\theta\,\mathbf{j}.$$

The true course of the boat is the direction that the boat actually travels, **w**, which in this case is due north. In other words, the **i**-component (or the horizontal component) of **w** must be 0. The overall velocity of the boat is the sum of the velocity of the river and the velocity of the boat: $\mathbf{w} = \mathbf{u} + \mathbf{v}$. Putting the pieces together, we have:

$\mathbf{w} = \mathbf{u} + \mathbf{v} = 5\mathbf{i} + 10\cos\theta\,\mathbf{i} + 10\sin\theta\,\mathbf{j} = (10\cos\theta + 5)\mathbf{i} + 10\sin\theta\,\mathbf{j}.$

Because the $\mathbf{i}$-component of $\mathbf{w}$ must be 0, we know that

$(10\cos\theta + 5) = 0.$

We can solve this equation for θ:

$(10\cos\theta + 5) = 0$

$10\cos\theta = -5$

$\cos\theta = -\frac{5}{10} = -\frac{1}{2}$

$\theta = 120°$

Vectors are useful in navigation. They can be used to calculate the speed and velocity of an airplane caught in a crosswind, or, as we saw in the previous example, to steer a boat across a river. Vectors can also be used to analyze the net force acting on a particle. A solid understanding of vectors is extremely important in analyzing problems in physics and engineering.

Lesson 8-1 Review

1. If $\mathbf{u} = \langle -5, 2\rangle$ and $\mathbf{v} = \langle 2, 7\rangle$ find the following:

    a. $2\mathbf{u} + \mathbf{v}$ b. $\mathbf{u} - 3\mathbf{v}$ c. $|3\mathbf{u}|$

2. Find the horizontal and vertical components of a vector whose magnitude is 7 and direction is $-\frac{2\pi}{3}$.

3. A river flows due south at a speed of 3 mph. A swimmer attempts to cross the river and heads due east at a rate of 2 mph relative to the water. Find the true velocity of the swimmer.

Lesson 8-2: The Dot Product

We have learned how to add and subtract vectors, and we have discussed scalar multiplication. It is natural to ask about the rules for multiplying or dividing two vectors. As it turns out, there are two ways that we can multiply vectors, but we cannot divide one vector by another vector.

We have been adding and multiplying real numbers for so long that it has become second nature to us. It is only when we construct a new mathematical object that we have to think about how we want to define addition and multiplication. The way that we define these operations will affect the mathematical universe that is created to contain these newly constructed objects. For example, when rational numbers, or fractions, were invented, rules had to be established for how to add and multiply fractions. In order to add two fractions together, we require that the denominators of the fractions be the same. As a result, addition of fractions usually involves finding a common denominator. The process of multiplying two fractions together involves multiplying the numerators together, and multiplying the denominators together. These rules were established and handed down from generation to generation. When we studied complex numbers, we also had to define how to add and multiply these new mathematical objects. We defined it in a way that seemed natural. When two complex numbers are added together, the real parts are added together and the imaginary parts are added together. Multiplication of binomial expressions served as the basis for multiplying two complex numbers together. This probably seemed like a reasonable procedure to use. As we just learned with vectors, we defined the addition of two vectors in a geometric way and an algebraic way. Both methods agreed, and the algebraic method (of adding the components of the vectors together) seemed natural. Now that we are on the topic of multiplying two vectors together, we need to put some thought into how to define this procedure.

The first thing we need to think about when we establish the rules for multiplying two vectors together is the type of object we want our answer to be. This may be something that you are not used to thinking about. For example, when we multiply two real numbers together, our answer is a real number. When we multiply two fractions together, our answer is a fraction. Well, our answer may actually be an integer, but because integers can be thought of as rational numbers, this possibility does not bother us. When we multiply two complex numbers together, our answer is a complex number. Again, there are complex numbers whose product is actually a real number, such as $(2 + i)(2 - i)$, but because we can think of a real number as a complex number whose imaginary part is 0, this possibility does not bother us. Vectors are a different matter. When we multiply two vectors together, we need to decide whether our answer should be a scalar or a vector. Scalars are *not* special cases of vectors, so the form that our answer takes will make a difference in this case. In fact, there are two ways to multiply two vectors together. With one method, the result is a scalar, and with the other method the result is a vector. We will examine one approach in this lesson, and a different definition of multiplication in the next lesson.

The first method for multiplying two vectors is called the **dot product**. The symbol for the dot product is a dot: $\cdot$. The dot product combines two vectors and the result is a scalar. If $v_1 = \langle a_1, b_1 \rangle$ and $v_2 = \langle a_2, b_2 \rangle$, then their dot product, denoted by $v_1 \cdot v_2$ is defined as:

$$v_1 \cdot v_2 = a_1 \cdot a_2 + b_1 \cdot b_2$$

From this equation, we see that to find the dot product of two vectors, multiply the corresponding components and add the results. The dot product is sometimes called the **inner product** of two vectors.

Example 1

If $v_1 = \langle 3, -2 \rangle$ and $v_2 = \langle 4, 1 \rangle$, find $v_1 \cdot v_2$.

Solution: Use the definition of the dot product:

$v_1 \cdot v_2 = 3 \cdot 4 + (-2) \cdot (1) = 12 - 2 = 10$

The dot product of a vector and itself is worth exploring. If $v = \langle a, b \rangle$, then $v \cdot v = \langle a, b \rangle \cdot \langle a, b \rangle = a^2 + b^2$. The expression $a^2 + b^2$ should look familiar. Remember that the modulus of v is $\sqrt{a^2 + b^2}$. The dot product of a vector with itself is the square of its modulus: $v \cdot v = |v|^2$. Some important properties of the dot product are summarized in the following table.

The dot product is commutative.	$u \cdot v = v \cdot u$
The dot product distributes over vector addition.	$(u + v) \cdot w = u \cdot w + v \cdot w$
The dot product is associative with respect to scalar multiplication.	$(au) \cdot v = a(u \cdot v) = u \cdot (a v)$
The dot product of a vector and itself is the square of its modulus.	$v \cdot v = |v|^2$
The dot product of a vector and the zero vector is 0.	$0 \cdot v = 0$

Trigonometry can be found throughout the development of vectors. Trigonometry can be used to find the horizontal and vertical components of a vector and the direction of a vector. When two vectors are in standard position, they form an angle. The angle between these two vectors can be determined using the Law of Cosines. The vectors **u**, **v**, and **u** + **v** form a triangle, and if $|\mathbf{u}|$, $|\mathbf{v}|$, and $|\mathbf{u}+\mathbf{v}|$ are known, then we can use the Law of Cosines to find the various interior angles of the triangle. The dot product is a much more direct way to find the angle formed by two vectors. The angle between the vectors **u** and **v** can be found using the formula:

$$\cos\theta = \frac{\mathbf{u}\bullet\mathbf{v}}{|\mathbf{u}||\mathbf{v}|}$$

Dot products and norms are fairly easy to calculate, if you know the components of the vectors **u** and **v**. The dot product provides an easy way to find the angle between two vectors. The formula $\cos\theta = \dfrac{\mathbf{u}\bullet\mathbf{v}}{|\mathbf{u}||\mathbf{v}|}$ can be derived using the Law of Cosines.

The sign of the dot product between two vectors indicates the size of the angle between them, θ. If $\mathbf{u}\cdot\mathbf{v}>0$, then θ is acute. If $\mathbf{u}\cdot\mathbf{v}<0$, then θ is obtuse. If $\mathbf{u}\cdot\mathbf{v}=0$, then $\theta=\frac{\pi}{2}$.

Example 2

Find the angle between the vectors $\mathbf{u}=\langle-2,5\rangle$ and $\mathbf{v}=\langle3,2\rangle$.

Solution: Use the formula $\cos\theta = \dfrac{\mathbf{u}\bullet\mathbf{v}}{|\mathbf{u}||\mathbf{v}|}$.

First, find $|\mathbf{u}|$, $|\mathbf{v}|$, and $\mathbf{u}\cdot\mathbf{v}$:

$$|\mathbf{u}|=\sqrt{(-2)^2+5^2}=\sqrt{29}$$

$$|\mathbf{v}|=\sqrt{3^2+2^2}=\sqrt{13}$$

$$\mathbf{u}\bullet\mathbf{v}=(-2)\cdot3+5\cdot2==-6+10=4$$

$$\cos\theta = \frac{\mathbf{u}\bullet\mathbf{v}}{|\mathbf{u}||\mathbf{v}|}=\frac{4}{\sqrt{377}}$$

$$\theta=\cos^{-1}\left(\frac{4}{\sqrt{377}}\right)\approx78.1^\circ$$

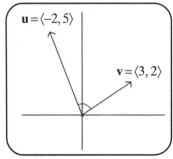

Figure 8.8.

The graphs of **u** and **v** are shown in Figure 8.8.

Example 3

Find the angle between the vectors **i** and **j**.

Solution: Remember that **i** and **j** are unit vectors, so $|\mathbf{i}| = 1$ and $|\mathbf{j}| = 1$. Also, $\mathbf{i} = \langle 1, 0 \rangle$ and $\mathbf{j} = \langle 0, 1 \rangle$, so $\mathbf{i} \cdot \mathbf{j} = \mathbf{0}$. From this, we see that $\cos\theta = \frac{0}{1} = 0$, which means that $\theta = 90°$.

The vectors **i** and **j** are perpendicular to each other, so it is not surprising that the angle between them is 90°. It does lead us to a new way of determining whether or not two vectors are perpendicular, or orthogonal to each other. Two non-zero vectors are **perpendicular**, or **orthogonal**, if the angle between them is 90°. Equivalently, two non-zero vectors are perpendicular, or orthogonal, if their dot product is 0.

Example 4

Are the following pairs of vectors orthogonal?

a. $\mathbf{u} = \langle -2, 3 \rangle$ and $\mathbf{v} = \langle -3, 2 \rangle$ b. $\mathbf{u} = \langle 1, 4 \rangle$ and $\mathbf{v} = \langle -4, 1 \rangle$

Solution: If $\mathbf{u} \cdot \mathbf{v} = \mathbf{0}$, then the vectors are orthogonal.

a. $\mathbf{u} \cdot \mathbf{v} = \langle -2, 3 \rangle \cdot \langle -3, 2 \rangle = 6 + 6 = 12$. Because $\mathbf{u} \cdot \mathbf{v} \neq 0$, the two vectors are not orthogonal.

b. $\mathbf{u} \cdot \mathbf{v} = \langle 1, 4 \rangle \cdot \langle -4, 1 \rangle = -4 + 4 = 0$. Because $\mathbf{u} \cdot \mathbf{v} = 0$, the two vectors are orthogonal.

Example 5

Find the value(s) of k so that the vectors $\mathbf{u} = \langle -3k, 1 \rangle$ and $\mathbf{v} = \langle 4, k^2 \rangle$ are orthogonal.

Solution: The two vectors will be orthogonal if their dot product is 0: $\mathbf{u} \cdot \mathbf{v} = \langle -3k, 1 \rangle \cdot \langle 4, k^2 \rangle = -12k + k^2$

From this we see that $\mathbf{u} \cdot \mathbf{v} = 0$ precisely when $-12k + k^2 = 0$:

$-12k + k^2 = 0$

$k(k - 12) = 0$

$k = 0$ or $k = 12$

The two vectors are orthogonal when $k = 0$ or $k = 12$.

In many applications it is of interest to write a given vector **u** as a sum of two vectors, $\mathbf{w}_1 + \mathbf{w}_2$, where $\mathbf{w}_1$ is parallel to a specified nonzero vector **a** and the $\mathbf{w}_2$ is perpendicular to **a**. This process involves *orthogonal projection*. In decomposing a vector this way, we will find out how much of **u** is applied in a direction that is parallel to the vector **a**, and how much of **u** is applied in a direction that is perpendicular to the vector **a**. The **orthogonal projection** of **u** onto **a**, or the vector component of **u** along **a**, is denoted by $\mathbf{w}_1 = proj_a \mathbf{u}$, and can be calculated using the formula:

$$\mathbf{w}_1 = proj_a \mathbf{u} = \left(\frac{\mathbf{u} \cdot \mathbf{a}}{|\mathbf{a}|^2} \right) \mathbf{a}$$

The quantity $\left(\frac{\mathbf{u} \cdot \mathbf{a}}{|\mathbf{a}|^2} \right)$ is a scalar: the dot product in the numerator is a scalar, and the magnitude, or norm, of a vector is also a scalar.

To decompose **u** into components relative to a vector **a**, we first find the orthogonal projection vector $\mathbf{w}_1$. Then $\mathbf{w}_2 = \mathbf{u} - \mathbf{w}_1$. From this second equation, we see that $\mathbf{u} = \mathbf{w}_1 + \mathbf{w}_2$.

Example 6

Let $\mathbf{u} = \langle 1, 3 \rangle$ and $\mathbf{a} = \langle 1, 1 \rangle$. Find the vector component of **u** along **a** and the vector component of **u** that is orthogonal to **a**.

Solution: The vector component of **u** along **a** is can be found

using the formula $\mathbf{w}_1 = proj_a \mathbf{u} = \left(\frac{\mathbf{u} \cdot \mathbf{a}}{|\mathbf{a}|^2} \right) \mathbf{a}$:

$$\mathbf{w}_1 = \left(\frac{\mathbf{u} \cdot \mathbf{a}}{|\mathbf{a}|^2} \right) \mathbf{a} = \left(\frac{4}{2} \right) \langle 1, 1 \rangle = \langle 2, 2 \rangle$$

The vector component of **u** that is *orthogonal* to **a** is just $\mathbf{w}_2 = \mathbf{u} - \mathbf{w}_1$:

$\mathbf{w}_2 = \mathbf{u} - \mathbf{w}_1 = \langle 1, 3 \rangle - \langle 2, 2 \rangle = \langle -1, 1 \rangle$

The graphs of **u**, **a**, $\mathbf{w}_1$, and $\mathbf{w}_2$ are shown in Figure 8.9.

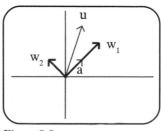

Figure 8.9.

In general, the **component** of **u** along **a** is given by the formula $\dfrac{\mathbf{u}\bullet\mathbf{a}}{|\mathbf{a}|}$.

As an example of decomposing a vector into orthogonal vectors, think back to when we found the horizontal and vertical components of a vector. Remember that the horizontal component of a vector $\mathbf{u} = \langle a, b\rangle$ is a. The component of **u** along **i** is:

$$\frac{\mathbf{u}\bullet\mathbf{i}}{|\mathbf{i}|} = \frac{\langle a,b\rangle\bullet\langle 1,0\rangle}{1} = a$$

Similarly, the vertical component of **u** along **j** is b. The horizontal and vertical components of a vector can be seen immediately; we do not have to use a formula for the component of a vector to find these components. The vectors **i** and **j** make this calculation very easy, which is one reason why these vectors are so important.

One application of the dot product involves the calculation of work done by a constant force in moving an object from one point to another. The work, W, done by a force, **F**, in coving an object along a vector, **D**, is $W = \mathbf{F} \cdot \mathbf{D}$. Work is a scalar quantity, and its units of measurement are either foot-pounds (ft-lbs) or joules.

Example 7

A constant force described by the vector $\mathbf{F} = \langle 3, 5\rangle$ moves an object from the point $(2, 6)$ to the point $(4, 8)$. Calculate the work done.

Solution: We know the vector for the force, but we need a vector to describe the displacement. We can use the initial and terminal points to find the components of the displacement vector:

$\mathbf{D} = \langle 4 - 2, 8 - 6\rangle = \langle 2, 2\rangle$.

The work done can be calculated using the formula $W = \mathbf{F} \cdot \mathbf{D}$:

$W = \mathbf{F} \cdot \mathbf{D} = \langle 3, 5\rangle \cdot \langle 2, 2\rangle = 6 + 10 = 16$

If the unit of force is pounds and the distance is measured in feet, then the work done is 16 ft-lb. If the unit of force is newtons and the distance is measured in meters, then the work done is 16 joules.

Example 8

A child pulls a wagon with a force of 40 pounds. How much work is done in moving the wagon 50 feet if the handle makes an angle of 30° with the ground?

Solution: The magnitude of the force is 40 pounds. We will need to decompose the force into its horizontal and vertical components. Remember that we can decompose a vector **v** into its horizontal and vertical components using the formula $v = |v|\cos\theta\mathbf{i} + |v|\sin\theta\mathbf{j}$.

From this, we see that $\mathbf{F} = \langle|\mathbf{F}|\cos\theta, |\mathbf{F}|\sin\theta\rangle$, where $|\mathbf{F}| = 40$ and $\theta = 30°$:

$$\mathbf{F} = \langle 40\cos 30°, 40\sin 30°\rangle = \langle 20\sqrt{3}, 20\rangle$$

The displacement vector only has a horizontal component: $\mathbf{D} = \langle 50, 0\rangle$. The work done is:

$$W = \mathbf{F}\cdot\mathbf{D} = \langle 20\sqrt{3}, 20\rangle\cdot\langle 50, 0\rangle = 100\sqrt{3} \approx 173$$

The work done is approximately 173 ft-lb.

Lesson 8-2 Review

1. Find the angle between the vectors $\mathbf{u} = \langle-2, 5\rangle$ and $\mathbf{v} = \langle 3, 2\rangle$.

2. Let $\mathbf{p} = \langle 2, k\rangle$ and $\mathbf{q} = \langle 3, 5\rangle$. Find k so that:
   a. **p** and **q** are parallel.   b. **p** and **q** are orthogonal.

3. A lawn mower is pushed a distance of 200 feet along a horizontal path by a constant force of 50 pounds. Compare the work done if the handle of the lawn mower is held at an angle of 30° to the work done if the handle of the lawn mower is held at an angle of 45°.

Lesson 8-3: The Cross Product

We have been positioning vectors in the coordinate plane, or two-dimensional space. Vectors in the plane are described by an ordered *pair* of real numbers. Vectors in three-dimensional space can be described by an ordered *triple* of real numbers. To construct a rectangular three-dimensional coordinate system, select a point, O, called the origin, and choose three mutually perpendicular lines, called the coordinate axes, that pass through the origin. Label these axes x, y, and z. Select a positive direction for each coordinate axis, and a unit of length for measuring distances. Each pair of coordinate axes determines a plane, called a coordinate plane. There are three coordinate planes: the xy-plane, the xz-plane,

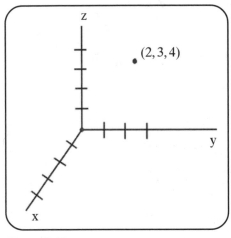

Figure 8.10.

and the yz-plane. To each point P in this space, we assign an ordered triple of numbers, (x, y, z), called the coordinates of P: x represents the perpendicular distance from P to the yz-plane, y represents the perpendicular distance from P to the xz-plane, and z represents the perpendicular distance from P to the xy-plane. Figure 8.10 shows a three-dimensional coordinate system with the point $(2, 3, 4)$.

The positive x-axis comes out of the page, the positive y-axis is to the right, and the positive z-axis is above.

Rectangular coordinate systems in three-dimensional space fall into two categories: left-handed and right-handed. A right-handed system has the property that if the right hand is held so that the curled fingers follow the rotation of the positive x-axis into the positive y-axis, the extended right thumb will point in the direction of the positive z-axis. This property is called the **right-hand rule**. Another way to describe the right-hand rule is to describe the motion of a regular screw. If, when looking at the head of a screw, you turn a screwdriver counter-clockwise, the screw should travel in the direction of the positive z-axis. Most machine parts are designed to obey the right-hand rule (as illustrated by the phrase "righty-tighty, lefty-loosy" which can help people decide how to turn a screw in order to tighten it or loosen it: turning a screw counter-clockwise loosens the screw, and turning it clockwise tightens it). Of course, there are machine parts that intentionally do not follow the right-hand rule. For example, propane tanks that are used to hold fuel for a gas fireplace use machine parts that follow the left-hand rule: their threads are backwards. For our purposes, we will only use a right-handed coordinate system.

Working with ordered triples is not much different than working with ordered pairs. Given the initial point (x_1, y_1, z_1) and terminal point (x_2, y_2, z_2) of a vector $\mathbf{v}$ in three-dimensional space, the components of the vector are found in the natural way: $\mathbf{v} = \langle x_2 - x_1, y_2 - y_1, z_2 - z_1 \rangle$. If $\mathbf{v} = \langle x, y, z \rangle$, then the magnitude of $\mathbf{v}$, which is still denoted by $|\mathbf{v}|$, is found using the formula $|\mathbf{v}| = \sqrt{x^2 + y^2 + z^2}$. The rules for adding and subtracting vectors follow the same methods as in Lesson 8-1. Addition and subtraction are done component-wise. The dot product between two vectors is also defined similarly for vectors in three-dimensions as it was in two-dimensions. If $\mathbf{v}_1 = \langle x_1, y_1, z_1 \rangle$ and $\mathbf{v}_2 = \langle x_2, y_2, z_2 \rangle$, then

$$\mathbf{v}_1 \cdot \mathbf{v}_2 = x_1 \cdot x_2 + y_1 \cdot y_2 + z_1 \cdot z_2.$$

All of these formulas for working with vectors in three-dimensional space should make sense. Adding one more component to a vector just involves incorporating one more component into the formulas for the magnitude of a vector and the dot product of two vectors.

Example 1

If $\mathbf{v}_1 = \langle 1, 2, 4 \rangle$ and $\mathbf{v}_2 = \langle -2, 0, 6 \rangle$, find the following:

a. $3\mathbf{v}_1 + 4\mathbf{v}_2$

b. $\mathbf{v}_1 \cdot \mathbf{v}_2$

c. The angle between $\mathbf{v}_1$ and $\mathbf{v}_2$.

Solution:

a. $3\mathbf{v}_1 + 4\mathbf{v}_2 = 3\langle 1, 2, 4 \rangle + 4\langle -2, 0, 6 \rangle = \langle 3, 6, 12 \rangle + \langle -8, 0, 24 \rangle$
$\phantom{3\mathbf{v}_1 + 4\mathbf{v}_2} = \langle -5, 6, 36 \rangle$

b. $\mathbf{v}_1 \cdot \mathbf{v}_2 = \langle 1, 2, 4 \rangle \cdot \langle -2, 0, 6 \rangle = -2 + 0 + 24 = 22$

c. To find the angle between $\mathbf{v}_1$ and $\mathbf{v}_2$, we will use the formula

$$\cos\theta = \frac{\mathbf{v}_1 \bullet \mathbf{v}_2}{|\mathbf{v}_1||\mathbf{v}_2|}: \qquad |\mathbf{v}_1| = \sqrt{1^2 + 2^2 + 4^2} = \sqrt{21}$$

$$|\mathbf{v}_2| = \sqrt{(-2)^2 + 0^2 + 6^2} = \sqrt{40}$$

$$\mathbf{v}_1 \cdot \mathbf{v}_2 = 22$$

$$\cos\theta = \frac{\mathbf{v}_1 \cdot \mathbf{v}_2}{|\mathbf{v}_1||\mathbf{v}_2|} = \frac{22}{\sqrt{21}\sqrt{40}} = \frac{11}{\sqrt{210}}$$

$$\theta = \cos^{-1}\frac{11}{\sqrt{210}} \approx 40.6°$$

As I mentioned in Lesson 8-2, there are two ways to multiply two vectors together. With the first method, called the dot product, the result of this multiplication is a scalar. With the second method, called the **cross product**, the result of this multiplication is a *vector*. We will define the cross product for vectors in three-dimensional space, and briefly discuss some of the properties and applications of the cross product. The cross product of two vectors, **u** and **v**, is only defined for vectors in three-dimensional space, and is denoted by **u** × **v**.

If $\mathbf{u} = \langle x_1, y_1, z_1 \rangle$ and $\mathbf{v} = \langle x_2, y_2, z_2 \rangle$, then **u** × **v** is defined as:

$$\mathbf{u} \times \mathbf{v} = \langle y_1 z_2 - y_2 z_1, -(x_1 z_2 - x_2 z_1), x_1 y_2 - x_2 y_1 \rangle$$

The formula for the cross product may seem long and difficult to memorize. Notice the symmetry of the formula. The calculation of the x component involves the difference of the products of the y and z components of **u** and **v**. Because *subtraction* is involved, the order in which we take the cross product matters: $\mathbf{u} \times \mathbf{v} \neq \mathbf{u} \times \mathbf{v}$. In other words, the cross product is not associative. This is just one of several basic algebraic properties of the cross product. Other algebraic properties of the cross product are summarized in the following table.

The cross-product of a vector and itself is the zero vector.	$\mathbf{v} \times \mathbf{v} = \mathbf{0}$
Changing the order in which two vectors are multiplied changes the sign of the resulting product.	$\mathbf{v} \times \mathbf{w} = -(\mathbf{w} \times \mathbf{v})$
Scalar multiplication associates with the cross-product.	$c(\mathbf{v} \times \mathbf{w}) = (c\mathbf{v}) \times \mathbf{w} = \mathbf{v} \times (c\mathbf{w})$
The cross-product distributes over vector addition.	$\mathbf{u} \times (\mathbf{v} + \mathbf{w}) = \mathbf{u} \times \mathbf{v} + \mathbf{u} \times \mathbf{w}$

Example 2

If $\mathbf{u} = \langle 1, 2, 4 \rangle$ and $\mathbf{v} = \langle -2, 0, 6 \rangle$, find $\mathbf{u} \times \mathbf{v}$.

Solution: Use the formula for the cross-product:

$$\mathbf{u} \times \mathbf{v} = \langle 2 \cdot 6 - 0 \cdot 4, -(1 \cdot 6 - (-2) \cdot 4), 1 \cdot 0 - (-2) \cdot 2 \rangle = \langle 12, -14, 4 \rangle$$

We can take a little time to compute the dot product of the vectors $\mathbf{u} = \langle 1, 2, 4 \rangle$ and $\mathbf{u} \times \mathbf{v} = \langle 12, -14, 4 \rangle$:

$$\mathbf{u} \cdot (\mathbf{u} \times \mathbf{v}) = \langle 1, 2, 4 \rangle \cdot \langle 12, -14, 4 \rangle = 12 - 28 + 16 = 0$$

Also, we can calculate the dot product of the vectors $\mathbf{v} = \langle -2, 0, 6 \rangle$ and $\mathbf{u} \times \mathbf{v} = \langle 12, -14, 4 \rangle$:

$$\mathbf{v} \cdot (\mathbf{u} \times \mathbf{v}) = \langle -2, 0, 6 \rangle \cdot \langle 12, -14, 4 \rangle = -24 + 24 = 0$$

From this we see that the vector $\mathbf{u} \times \mathbf{v}$ is orthogonal to both $\mathbf{u}$ and $\mathbf{v}$. This is an important property of the cross product. Given any two vectors $\mathbf{u}$ and $\mathbf{v}$, the easiest way to find a vector that is orthogonal to *both* $\mathbf{u}$ and $\mathbf{v}$ is to take their cross product. This is just one of many special geometric properties of the cross product. The geometric properties of the cross product are summarized in the following table.

$\mathbf{u} \times \mathbf{v}$ is orthogonal to both $\mathbf{u}$ and $\mathbf{v}$.	$\mathbf{u} \cdot (\mathbf{u} \times \mathbf{v}) = 0$ and $\mathbf{v} \cdot (\mathbf{u} \times \mathbf{v}) = 0$
The magnitude of $\mathbf{u} \times \mathbf{v}$ is related to the sine of the angle between $\mathbf{u}$ and $\mathbf{v}$.	$\|\mathbf{u} \times \mathbf{v}\| = \|\mathbf{u}\| \|\mathbf{v}\| \sin \theta$
The area of the parallelogram with (nonzero) sides $\mathbf{u}$ and $\mathbf{v}$ is equal to the magnitude of the cross-product of $\mathbf{u}$ and $\mathbf{v}$.	$\text{Area}_{\text{parallelogram}} = \|\mathbf{u} \times \mathbf{v}\|$
The cross-product of two parallel vectors is the zero vector.	$(\mathbf{v} \times \mathbf{u}) = 0$ if $\mathbf{u}$ is parallel to $\mathbf{v}$

One application of the cross product involves finding a vector that is orthogonal to a given plane. To find such a vector, first find two vectors that lie in the plane. The cross product of these two vectors will be orthogonal to the plane that contains the two vectors.

Example 3

The consecutive vertices of a parallelogram are $(-2, 3, 0)$, $(0, 0, 0)$, $(1, 2, 3)$, and $(-1, 5, 3)$. Find the area of the parallelogram.

Solution: To find the area of the parallelogram, we need to find two vectors that represent two sides of the parallelogram that share a common vertex. From there, we can either find the sine of the angle between the two sides and use the equation $\text{Area}_{\text{parallelogram}} = |\mathbf{u} \times \mathbf{v}| = |\mathbf{u}| |\mathbf{v}| \sin \theta$, or we can evaluate the cross product directly. The points given in the problem are the consecutive points of a parallelogram, meaning that any two consecutive points in the list form one of the sides of the parallelogram. To find two sides of the parallelogram that share a common vertex, take any three consecutive points. The point in the middle will be the common vertex. Use that point as the initial point for each vector. In this problem, use the consecutive points $(-2, 3, 0)$, $(0, 0, 0)$, and $(1, 2, 3)$. The point $(0, 0, 0)$ will be the initial point for the two vectors. Use each of the remaining two points as a terminal end of a vector: the two terminal points will be $(-2, 3, 0)$ and $(1, 2, 3)$. One side of the parallelogram is formed by the vector with terminal point $(1, 2, 3)$ and initial point $(0, 0, 0)$. To find this vector, subtract the components of the initial point from the components of the terminal point: $\langle 1 - 0, 2 - 0, 3 - 0 \rangle = \langle 1, 2, 3 \rangle$. Another side of the parallelogram is formed by the vector with terminal point $(-2, 3, 0)$ and initial point $(0, 0, 0)$. To find this vector, subtract the components of the initial point from the components of the terminal point: $\langle -2 - 0, 3 - 0, 0 - 0 \rangle = \langle -2, 3, 0 \rangle$. Two sides of the parallelogram that share a common vertex are represented by the vectors $\langle 1, 2, 3 \rangle$ and $\langle -2, 3, 0 \rangle$. Evaluating their cross product, we have:

$$\langle 1, 2, 3 \rangle \times \langle -2, 3, 0 \rangle = \langle 2 \cdot 0 - 3 \cdot 3, -(1 \cdot 0 - (-2) \cdot 3), 1 \cdot 3 - (-2) \cdot 2 \rangle$$
$$= \langle -9, -6, 7 \rangle$$

The area of the parallelogram will be the magnitude of this vector:

$$\left| \langle 1, 2, 3 \rangle \times \langle -2, 3, 0 \rangle \right| = \sqrt{(-9)^2 + (-6)^2 + 7^2} = \sqrt{166}$$

As a means of checking your work, you can always make sure that any intermediate results are correct. For example, once the cross product was found, check to make sure that it is orthogonal to the given vectors: if both of the dot products are 0, then you can feel more confident about your answer.

Example 4

Find a unit vector that is orthogonal to the plane containing the vectors $\langle 1, 2, -1 \rangle$ and $\langle -2, 1, 0 \rangle$.

Solution: First, find a vector that is orthogonal to the two given vectors:

$$\langle 1, 2, -1 \rangle \times \langle -2, 1, 0 \rangle =$$
$$\langle 2 \cdot 0 - 1 \cdot (-1), -(1 \cdot 0 - (-2) \cdot 1), 1 \cdot 1 - (-2) \cdot 2 \rangle = \langle 1, 2, 5 \rangle$$

Next, normalize the vector by dividing by its magnitude:

$$|\langle 1, 2, 5 \rangle| = \sqrt{1^2 + 2^2 + 5^2} = \sqrt{30}.$$

The vector $\left\langle \frac{1}{\sqrt{30}}, \frac{2}{\sqrt{30}}, \frac{5}{\sqrt{30}} \right\rangle$ is a unit vector that is orthogonal to the plane containing the vectors $\langle 1, 2, -1 \rangle$ and $\langle -2, 1, 0 \rangle$.

There are many more applications of vectors, dot products and cross products. Linear algebra is an area of mathematics that focuses on analyzing the properties of vectors in general.

Lesson 8-3 Review

1. If $v_1 = \langle 3, -1, 0 \rangle$ and $v_2 = \langle 1, -4, 2 \rangle$, find the following:

   a. $3v_1 + 4v_2$

   b. $v_1 \cdot v_2$

   c. The angle between v_1 and v_2.

   d. A unit vector that is orthogonal to the plane containing the two vectors

2. If $u = \langle -1, 2, 1 \rangle$ and $v = \langle 0, 2, 3 \rangle$, find $u \times v$.

Answer Key

Lesson 8-1 Review

1. a. $2\mathbf{u} + \mathbf{v} = \langle -8, 11 \rangle$ b. $\mathbf{u} - 3\mathbf{v} = \langle -11, -19 \rangle$ c. $|3\mathbf{u}| = 3\sqrt{29}$

2. The horizontal component is $-\frac{7}{2}$, and the vertical component is $\frac{7\sqrt{3}}{2}$.

3. The true velocity of the swimmer is the vector $\langle 2, -3 \rangle$.

Lesson 8-2 Review

1. $\mathbf{u} \cdot \mathbf{v} = 4$, and $\cos\theta = \dfrac{4}{\sqrt{29}\sqrt{13}} = \dfrac{4}{\sqrt{377}}$, $\theta = \cos^{-1}\dfrac{4}{\sqrt{377}} \approx 78.1°$

2. Let $\mathbf{p} = \langle 2, k \rangle$ and $\mathbf{q} = \langle 3, 5 \rangle$. Find k so that:
    a. If $\mathbf{p}$ and $\mathbf{q}$ are parallel if there is a scalar c so that $\mathbf{p}\,c\mathbf{p}: \langle 2, k \rangle = c\langle 3, 5 \rangle$,
    which means that $2 = 3c$, or $c = \frac{2}{3}$. Then $k = 5c$, so $k = 5 \cdot \frac{2}{3} = \frac{10}{3}$.

    b. If $\mathbf{p}$ and $\mathbf{q}$ are orthogonal, then $\mathbf{p} \cdot \mathbf{q} = 0$: $\mathbf{p} \cdot \mathbf{q} = 6 + 5k = 0$, so $k = -\frac{6}{5}$.

3. The work if the handle is at an angle of 30° is
    $W_{30} = \mathbf{F} \cdot \mathbf{D} = \langle 50\cos 30°, 50\sin 30° \rangle \cdot \langle 200, 0 \rangle = 8{,}660$ ft-lb
    The work if the handle is at an angle of 45° is
    $W_{45} = \mathbf{F} \cdot \mathbf{D} = \langle 50\cos 45°, 50\sin 45° \rangle \cdot \langle 200, 0 \rangle = 7{,}071$ ft-lb
    $W_{30} > W_{45}$

Lesson 8-3 Review

1. a. $3\mathbf{v}_1 + 4\mathbf{v}_2 = \langle 13, -19, 8 \rangle$
    b. $\mathbf{v}_1 \cdot \mathbf{v}_2 = 7$

    c. $\cos\theta = \dfrac{7}{\sqrt{10}\sqrt{21}} = \dfrac{7}{\sqrt{210}}$, $\theta = \cos^{-1}\dfrac{7}{\sqrt{210}} \approx 50.7°$

    d. $\mathbf{v}_1 \times \mathbf{v}_2 = \langle -2, -6, -11 \rangle$, and a unit vector that satisfies the conditions is
    $\left\langle \dfrac{-2}{\sqrt{161}}, \dfrac{-6}{\sqrt{161}}, \dfrac{-11}{\sqrt{161}} \right\rangle$

2. $\mathbf{u} \times \mathbf{v} = \langle 4, 3, -2 \rangle$

Analytic Geometry

The Cartesian coordinate system enables mathematicians to solve geometry problems algebraically. With this coordinate system, we are also able to take geometric concepts and describe them with algebraic equations. For example, from a geometric perspective, a circle is the set of all points that are a fixed distance r from a given point C, called the center of the circle. If the coordinates of the circle are (h, k), then the formula for the circle is $(x - h)^2 + (y - k)^2 = r^2$. Trigonometry is based on this algebraic description of a circle. There are other geometric concepts that can be described using algebraic equations. In this chapter, we will use the geometric descriptions of parabolas, ellipses, and hyperbolas to discover their algebraic equivalents.

The geometric definitions of a circle, a parabola, an ellipse, and a hyperbola are similar in that they all involve the distance between a specific point, called a **focus** (or points, called **foci**), and some other object. Circles, parabolas, ellipses, and hyperbolas are called **conics**, or conic sections, because their shapes can be generated by intersecting a plane and a cone, as shown in Figure 9.1.

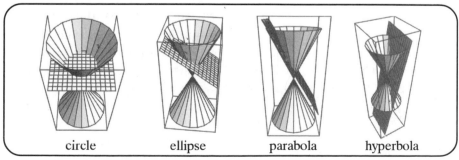

circle   ellipse   parabola   hyperbola

Figure 9.1.

These four curves all result from slicing a pair of cones with a plane. If a plane slices one of the cones parallel to its base, the resulting conic is a circle. If the cones are sliced at an angle, the result is an elongated circle, or an ellipse. The more we tilt the slice, the more elongated the ellipse becomes. When the slice is parallel to the side of the cone, the curve is no longer closed, and the result is a parabola. If the angle of the slice is tilted further, the plane intersects both cones and a hyperbola is created.

Each conic will have its own terminology. There are several important features of the various conics that will enable us to sketch their graphs easily. After we examine each conic in detail, we will see how trigonometry and polar coordinates can help us define all three conics in the same way.

Lesson 9-1: Parabolas

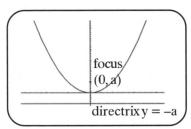

Figure 9.2.

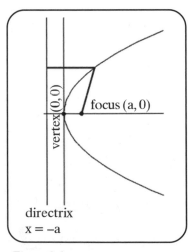

Figure 9.3.

From algebra, we know that the shape of the graph of a quadratic equation is called a parabola. Geometrically, a parabola is defined as the collection of all points P in the plane that are the same distance from a fixed point F as they are from a fixed line D. The point F is called the **focus** of the parabola, and the line D is called the **directrix**.

Figure 9.2 shows a parabola, its focus, and its directrix. The line that passes through the focus F and is perpendicular to the directrix D is called the **axis of symmetry** of the parabola. The point where the axis of symmetry and the parabola intersect is called the **vertex** of the parabola.

Any point in the Cartesian coordinate system can serve as the focus of a parabola, and any line in the Cartesian coordinate system can serve as the directrix. We will use the distance formula to derive a formula for a parabola. In this lesson, we will consider the special case where the directrix is parallel to one of the coordinate axes.

Suppose that the vertex V is located at the origin, and that the focus F is located on the positive x-axis, at the point $(a, 0)$, where $a > 0$, as shown in Figure 9.3. Then the directrix must be the line $x = -a$.

If $P = (x, y)$ is any point on the parabola, then the distance between P and the focus must equal the distance between P and the directrix. Now, the distance between P and the focus is $\sqrt{(x-a)^2 + (y-0)^2}$, or $\sqrt{(x-a)^2 + y^2}$. The distance between P and the directrix is $|x + a|$. We can set these two distances equal to each other to find the equation of this parabola:

$$\sqrt{(x-a)^2 + y^2} = |x + a|$$

Square both sides of this equation $(x - a)^2 + y^2 = (x + a)^2$

Expand $(x - a)^2$ and $(x + a)^2$ $x^2 - 2ax + a^2 + y^2 = x^2 + 2ax + a^2$

Simplify $y^2 = 4ax$

The graph of the parabola $y^2 = 4ax$ is shown in Figure 9.4. The parabola lies entirely in Quadrant I and Quadrant IV; the x-coordinate of every point on the parabola must be positive, because $a > 0$ and $y^2 > 0$.

If we keep the vertex of the parabola at the origin, we can also place the focus F on the positive y-axis, the negative x-axis, or the negative y-axis. The characteristics of the general form of a parabola are given in the table shown on page 194, and their graphs are shown in Figure 9.5.

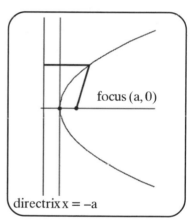

focus $(a, 0)$

directrix $x = -a$

Figure 9.4.

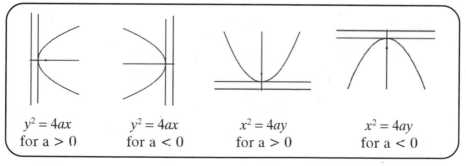

| $y^2 = 4ax$ | $y^2 = 4ax$ | $x^2 = 4ay$ | $x^2 = 4ay$ |
| for $a > 0$ | for $a < 0$ | for $a > 0$ | for $a < 0$ |

Figure 9.5.

Equation	$y^2 = ax$	$x^2 = ay$
Focus	$\left(\frac{a}{4}, 0\right)$	$\left(0, \frac{a}{4}\right)$
Directrix	$x = -\frac{a}{4}$	$y = -\frac{a}{4}$
Axis of Symmetry	x-axis	y-axis
Opens	Right if $a > 0$; Left if $a < 0$	Up if $a > 0$; Down if $a < 0$
Quadrants	I and IV if $a > 0$, II and III if $a < 0$	I and II if $a > 0$, III and IV if $a < 0$
Example	$y^2 = x$	$x^2 = y$

Example 1

For the following equations, determine the focus, the directrix, the axis of symmetry, and whether the parabola opens to the left, the right, up, or down:

a. $y^2 = 12x$

b. $x^2 = -6y$

Solution:

a. $y^2 = 12x$: This parabola is of the form $y^2 = ax$, where $a = 12$. The focus is $\left(\frac{a}{4}, 0\right)$, or $(3, 0)$. The directrix is $x = -3$, the axis of symmetry is the x-axis, and it opens to the right. The graph of $y^2 = 12x$ is shown in Figure 9.6.

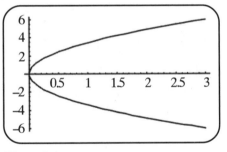

Figure 9.6.

b. $x^2 = -6y$: This parabola is of the form $x^2 = ay$, where $a = -6$. The focus is $\left(0, -\frac{6}{4}\right)$, or $\left(0, -\frac{3}{2}\right)$, the directrix is $y = \frac{3}{2}$, the axis of symmetry is the y-axis, and it opens down. The graph of $x^2 = -6y$ is shown in Figure 9.7.

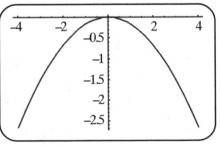

Figure 9.7.

Example 2

Find the equation of a parabola whose vertex is the point $(0, 0)$, axis of symmetry is the x-axis, and graph contains the point $(2, 3)$.

Solution: The axis of symmetry is the x-axis, so the general form of the parabola is $y^2 = ax$. We can substitute the point $(2, 3)$ into the equation of the parabola to solve for a:

$y^2 = ax$

$3^2 = a \cdot 2$

$a = \frac{9}{2}$

The equation of the parabola is $y^2 = \frac{9}{2}x$.

There is no requirement that the vertex of a parabola has to be located at the origin. There is also no requirement that the directrix and the axis of symmetry have to be either parallel or perpendicular to the coordinate axes. We will discuss the former case now, and the latter case in Lesson 9-4.

A parabola can be translated horizontally or vertically. When a parabola is translated, every point on the parabola is translated by the same amount. In particular, the vertex of the parabola is translated. The equations $y^2 = ax$ and $x^2 = ay$ describe parabolas whose vertex is $(0, 0)$. We could write these equations as $(y - 0)^2 = a(x - 0)$ and $(x - 0)^2 = a(y - 0)$, if we wanted to explicitly include the location of the vertex of the parabola. Remember that shifting a function vertically involves adding a constant to y, and shifting a function horizontally involves adding a constant to x. In general, the equation of a parabola whose vertex is the point (h, k) can be written in one of the following two forms:

$(y - k)^2 = a(x - h)$ or $(x - h)^2 = a(y - k)$,

depending on whether the axis of symmetry is parallel to the x-axis or the y-axis. It may be helpful to realize that the vertex and the focus of a parabola must lie on the axis of symmetry, regardless of how the parabola is oriented in the plane. The table following on page 196 describes the four different parabolas whose vertex is the point (h, k).

Notice that the distance between the vertex and the focus is $\left|\frac{a}{4}\right|$. This relationship may be used to find $|a|$.

Equation	$(y-k)^2 = a(x-h)$	$(x-h)^2 = a(y-k)$
Vertex	(h,k)	(h,k)
Focus	$\left(h+\frac{a}{4},k\right)$	$\left(h,k+\frac{a}{4}\right)$
Directrix	$x=h-\frac{a}{4}$	$y=k-\frac{a}{4}$
Axis of Symmetry	$y=k$	$x=h$
Opens	Right if $a>0$; Left if $a<0$	Up if $a>0$; Down if $a<0$
Example	$(y-2)^2 = 4(x-3)$	$(x-2)^2 = 4(y-3)$

Example 3

Find the equation of the parabola whose vertex is located at the point $(-2, 3)$ and whose focus is the point $(0, 3)$.

Solution: Think of yourself as a detective when solving these problems. The location of the vertex and the focus are clues to a puzzle. Use the preceding table to help put the pieces together. To begin with, the vertex and the focus must lie on the axis of symmetry. The equation of the line that passes through the points $(-2, 3)$ and $(0, 3)$ is $y = 3$. So the axis of symmetry is a horizontal line, meaning that the directrix is a vertical line. The distance between the vertex and the focus is 2, so

$\left|\frac{a}{4}\right| = 2$, and $|a| = 8$.

The focus lies to the right of the vertex, so the parabola must open to the right, meaning that $a > 0$. The equation of this parabola must be of the form $(y - k)^2 = a(x - h)$, and we know a, h, and k:
$(y - 3)^2 = 8(x - (-2))$, or $(y - 3)^2 = 8(x + 2)$

Example 4

Find the location of the vertex, the focus, and the directrix of the parabola $x^2 + 4x + 4y = 0$.

Solution: The equation needs to be in the form

$(x-h)^2 = a(y-k)$. In order to do this, we need to complete the square for the terms that involve x: $\qquad x^2 + 4x + 4y = 0$

Divide the linear coefficient of x by 2 and square the result. Add this number to both sides of the equation $\quad x^2 + 4x + 4 + 4y = 0 + 4$

Factor the quadratic expression involving x $\;(x+2)^2 + 4y = 4$

Subtract $4y$ from both sides $\qquad\qquad (x+2)^2 = -4y + 4$

Factor -4 from both terms on the right $\qquad (x+2)^2 = -4(y-1)$

The vertex is the point $(-2, 1)$.

From the equation of the parabola, we see that $h = -2$, $k = 1$, and $a = -4$. To find the focus, substitute in the values for h, k, and a

into the formula $\left(h, k + \frac{a}{4}\right)$ and simplify: $\qquad \left(-2, 1 + \left(\frac{-4}{4}\right)\right)$

$$(-2, 0)$$

To find the directrix, substitute in the values for h, k, and a into

the formula $y = k - \frac{a}{4}$ and simplify: $\qquad y = 1 - \left(\frac{-4}{4}\right) = 2$

The vertex is the point $(-2, 1)$, the focus is the point $(-2, 0)$, and the directrix is the line $y = 2$.

Objects that are parabola-shaped have the ability to collect low intensity signals from sources that are very far away. Light or sound that hit a parabolic surface are reflected, or channeled, to the focus. The intensity of the signal is much stronger at the focus than it is anywhere on the surface of the parabola. Satellite dishes, searchlights, and automobile headlights are a few examples of objects that make use of the properties of a parabola in their design. Science museums often have two parabolic dishes set up on opposite sides of a large room, and a metal circle located at the focus of each of the dishes. If one person whispers near one focus, the sound will be audible to someone listening at the other focus.

Example 5

A satellite dish is shaped like a paraboloid of revolution. The signals from a satellite will strike the surface of the dish and be collected at the receiver, which is located at the focus of the paraboloid. If the dish is 12 feet across at its opening and 6 feet deep at its center, where should the receiver be placed?

Solution: In this problem, it does not matter which way the satellite dish is oriented. It also does not matter where the satellite dish is located in the coordinate plane. The only thing that matters is the relative distances for the dimensions of the satellite dish. We can orient the satellite dish so that it points upward,

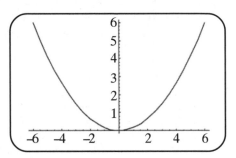

Figure 9.8.

with the vertex of the dish at the origin and the focus located along the positive y-axis, as shown in Figure 9.8. The dish is 12 feet across, and 6 feet deep, which means that the parabola will pass through the points (6, 6) and (–6, 6). With this orientation, the equation of the parabola will be $x^2 = ay$, and its focus will be the point $\left(0, \frac{a}{4}\right)$. Because (6, 6) is a point on the parabola, we can substitute this information into the equation for the parabola and solve for a:

$x^2 = ay$

$6^2 = a \cdot 6$

$a = \frac{36}{6} = 6$

The focus is the point $\left(0, \frac{a}{4}\right)$, or $\left(0, \frac{3}{2}\right)$. The receiver should be placed 1.5 feet from the base of the dish, along its axis of symmetry.

Lesson 9-1 Review

1. For the following equations, determine the focus, the directrix, the axis of symmetry, and whether the parabola opens to the left, the right, up or down:

    a. $y^2 = -10x$ 　　　　　　　　b. $x^2 = 4y$

2. Find the equation of a parabola whose vertex is the point $(0, 0)$, axis of symmetry is the x-axis, and graph contains the point $(4, 3)$.

3. Find the equation of the parabola whose vertex is located at the point $(3, -2)$ and whose focus is the point $(3, 4)$.

4. Find the location of the vertex, the focus, and the directrix of the parabola $y^2 + 6y + x = 0$.

5. A satellite dish is shaped like a paraboloid of revolution. The signals from a satellite will strike the surface of the dish and be collected at the receiver, which is located at the focus of the paraboloid. If the dish is 10 feet across at its opening and 4 feet deep at its center, where should the receiver be placed?

Lesson 9-2: Ellipses

An ellipse looks like an elongated, or stretched, circle. Orbits of planets, moons, and satellites follow an elliptical path. An **ellipse** is defined as the collection of all points in the plane the sum of whose distances from two fixed points F_1 and F_2 is a constant. The two fixed points are called the **foci** of the ellipse.

From this geometric definition of an ellipse, we can devise a physical way to draw an ellipse. Attach the two endpoints of a string to a piece of cardboard using thumbtacks. The length of the string is the constant sum referred to in the definition of the ellipse, and the two endpoints represent the foci of the ellipse. Use the point of a pencil to tighten the string and move the pencil around the foci, keeping the string tight at all times. The pencil will trace out an ellipse.

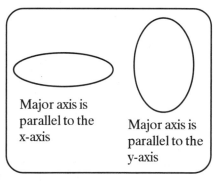

Major axis is parallel to the x-axis

Major axis is parallel to the y-axis

Figure 9.9.

If the string is only slightly longer than the distance between the foci, then the ellipse that is traced out will be elongated in shape. If the foci are close together relative to the length of the string, then the ellipse will be almost circular. Figure 9.9 shows ellipses with different degrees of elongation.

If F_1 and F_2 are the foci of an ellipse, then the line containing F_1 and F_2 is called the **major axis**. The midpoint of the line segment $\overline{F_1 F_2}$ is called the **center** of the ellipse. The line that passes through the center of the ellipse and is perpendicular to the major axis is called the **minor axis**. The points where the ellipse intersects the major axis are the **vertices** of the ellipse. An ellipse is symmetric with respect to its major axes, its minor axes, and its center.

The general equation of an ellipse whose center is located at the origin is given by the equation:

$$\frac{x^2}{a^2} + \frac{y^2}{b^2} = 1$$

If $0 < b < a$, then the major axis is the x-axis, the vertices are located at $(\pm a, 0)$, the horizontal length of the ellipse is $2a$ and the vertical length of the ellipse is $2b$. The foci are located at $(\pm c, 0)$, where c satisfies the equation $c^2 = a^2 - b^2$.

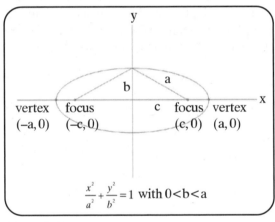

$$\frac{x^2}{a^2} + \frac{y^2}{b^2} = 1 \text{ with } 0 < b < a$$

Figure 9.10.

Most of these details can be determined from the equation $\frac{x^2}{a^2} + \frac{y^2}{b^2} = 1$, or from the graph of the ellipse, such as the one shown in Figure 9.10, and should not be memorized.

The points where the ellipse intersects the coordinate axes are just the intercepts of the equation $\frac{x^2}{a^2} + \frac{y^2}{b^2} = 1$.

Remember that the x-intercepts are found by evaluating the formula when $y = 0$, and the y-intercepts are found by evaluating the formula when $x = 0$. It is pretty easy to see that the x-intercepts are the solutions to the equation $\frac{x^2}{a^2} = 1$, and the y-intercepts are the solutions to the equation $\frac{y^2}{b^2} = 1$.

The location of the foci can be determined using the Pythagorean Theorem. Notice that we can construct a right triangle within the ellipse. The lengths of the two legs of the right triangle are b and c, and the length of the hypotenuse is a. From the Pythagorean Theorem, we see that $b^2 + c^2 = a^2$. The length of the major axis is the distance between the two vertices of the ellipse.

If, on the other hand, $0 < a < b$, the role of a and the role of b switch. The major axis is the y-axis, and the vertices are located at $(0, \pm b)$. The horizontal length of the ellipse is still $2a$ and the vertical length of the ellipse is still $2b$. The foci are located at $(0, \pm c)$, where c satisfies the equation $c^2 = b^2 - a^2$.

Most of these details can be determined from the equation $\frac{x^2}{a^2} + \frac{y^2}{b^2} = 1$, or from the graph of the ellipse, such as the one shown in Figure 9.11, and should not be memorized.

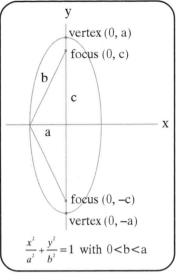

Figure 9.11.

The key to working with ellipses is to remember that the variable with the larger denominator wins. The variable with the larger denominator serves as the major axis, and gets to have the foci and the vertices. A little common sense leads you to the equation $c^2 = |a^2 - b^2|$ to find the non-zero coordinates of the foci. If $a > b$ then the foci are on the x-axis, and if $a < b$, the foci are on the y-axis. To sketch the graph of an ellipse, find the x-intercepts and the y-intercepts and sketch an ellipse that passes through those points. The longer axis is the major axis, and the smaller axis is the minor axis. The **length of the major axis** is the distance between the two vertices of the ellipse.

The **eccentricity** of an ellipse is defined as the ratio of the distance between the foci of the ellipse to the distance between the two vertices of the ellipse. Remember that the foci and the vertices of an ellipse lie on the major axis. If the foci are located at $(\pm c, 0)$ and the vertices are located at $(\pm a, 0)$, then the eccentricity of the ellipse, e, is given by the formula:

$$e = \frac{c}{a}$$

If the foci are located at $(0, \pm c)$ and the vertices are located at $(0, \pm b)$, then the eccentricity of the ellipse, e, is given by the formula:

$$e = \frac{c}{b}$$

The eccentricity of an ellipse is a number between 0 and 1, and its value indicates how elongated the ellipse is. The greater the elongation, the more oval the ellipse, or the more the ellipse deviates from a circle. If e is close to 1, then c is close to a, which means that b is small. In this case, the ellipse is elongated in shape. If e is close to 0, then the ellipse is close to a circle in shape.

Example 1

Sketch the graph of the ellipse given by the formula $\dfrac{x^2}{9} + \dfrac{y^2}{25} = 1$.

Find the vertices, the foci, and the eccentricity of the ellipse.

Solution: The denominator of the term that involves y is larger, so the major axis will be the y-axis, and the minor axis will be the x-axis. From the formula for the ellipse, we have that $a = 3$ and $b = 5$. The y-coordinate of the foci are the solutions to the equation $c^2 = b^2 - a^2 = 5^2 - 3^2 = 25 - 9 = 16$. The foci are located at $(0, \pm 4)$. The vertices are located at $(0, \pm 5)$.

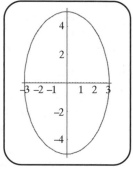

The eccentricity is $e = \dfrac{c}{b} = \dfrac{4}{5}$.

Figure 9.12.

The sketch of the ellipse is shown in Figure 9.12.

Example 2

The vertices of an ellipse are $(0, \pm 4)$ and the foci are $(0, \pm 2)$. Find the equation of the ellipse and its eccentricity. Sketch a graph of the ellipse.

Solution: The eccentricity is $e = \dfrac{2}{4} = \dfrac{1}{2}$. The vertices and the foci of the ellipse are along the y-axis, so the major axis is the y-axis. The foci can be found using the equation $c^2 = b^2 - a^2$. We know the values of b and c, so we can find a^2:

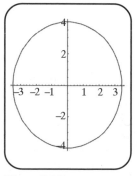

Figure 9.13.

$$c^2 = b^2 - a^2$$
$$2^2 = 4^2 - a^2$$
$$4 = 16 - a^2$$
$$a^2 = 12$$

The equation of the ellipse is $\dfrac{x^2}{12} + \dfrac{y^2}{16} = 1$.

The sketch of the ellipse is shown in Figure 9.13.

Example 3

The foci of an ellipse are $(0, \pm 8)$, and the eccentricity of the ellipse is $e = \frac{4}{5}$. Find the equation of the ellipse and its vertices. Sketch a graph of the ellipse.

Solution: The foci of the ellipse lie on the y-axis, so the y-axis is the major axis. We can use the foci of the ellipse and the eccentricity to find the nonzero coordinate of one of the vertices:

$$e = \frac{c}{b}$$

$$\frac{4}{5} = \frac{8}{b}$$

$$b = 10$$

The vertices of the ellipse are $(0, \pm 10)$. We can use the equation $c^2 = b^2 - a^2$ to find a^2:

$$c^2 = b^2 - a^2$$
$$8^2 = 10^2 - a^2$$
$$64 = 100 - a^2$$
$$a^2 = 36$$

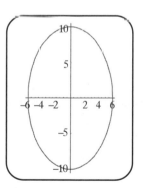

Figure 9.14.

The equation of the ellipse is $\dfrac{x^2}{36} + \dfrac{y^2}{100} = 1$.

The sketch of the ellipse is shown in Figure 9.14.

An ellipse can be translated horizontally or vertically. When an ellipse is translated, every point on the ellipse is translated by the same amount. In particular, the center, the foci, and the vertices of the ellipse is translated. An ellipse whose center is (0, 0) is described by the equation $\frac{x^2}{a^2} + \frac{y^2}{b^2} = 1$. We could write this equation as $\frac{(x-0)^2}{a^2} + \frac{(y-0)^2}{b^2} = 1$, if we wanted to explicitly include the location of the center of the ellipse. In general, the equation of an ellipse whose center is the point (h, k) can be written as:

$$\frac{(x-h)^2}{a^2} + \frac{(y-k)^2}{b^2} = 1$$

It may be helpful to realize that the vertices and the foci of an ellipse must lie on the major axis, regardless of how the ellipse is oriented in the plane. The values of a, b, and c no longer represent the nonzero coordinates of the vertices or the foci. Rather, these values represent the distance between the center and either a vertex or a focus. In particular, the value of a (or b, depending on which one is larger) represents the distance between the center of the ellipse and one vertex (where the ellipse intersects the major axis). The value of c represents the distance between the center of the ellipse and one of the foci. The value of b (or a, depending on which one is smaller) represents the distance between the center of the ellipse and the point where the ellipse intersects the minor axis. The following table summarizes the information for an ellipse with center (h, k).

Equation for the ellipse	$\frac{(x-h)^2}{a^2} + \frac{(y-k)^2}{b^2} = 1$, with $0 < b < a$	$\frac{(x-h)^2}{a^2} + \frac{(y-k)^2}{b^2} = 1$, with $0 < a < b$
Major axis	Parallel to the x-axis	Parallel to the y-axis
Vertices	$(h \pm a, k)$	$(h, k \pm b)$
Foci	$(h \pm c, k)$	$(h, k \pm c)$
Eccentricity	$e = \frac{c}{a}$	$e = \frac{c}{b}$

Example 4

Find the equation of the ellipse with one focus at $(4, 8)$ and vertices located at $(4, 3)$ and $(4, 9)$.

Solution: The focus and vertices must lie on the same line. In this case, all three points lie on the line $x = 4$. The major axis is parallel to the y-axis, so the equation of the ellipse will be

$\dfrac{(x-h)^2}{a^2} + \dfrac{(y-k)^2}{b^2} = 1$, where $0 < a < b$. We will need to find the

center of the ellipse, which will be the midpoint of the line segment connecting the two vertices. The midpoint of the line segment whose endpoints are $(4, 3)$ and $(4, 9)$ is $(4, 6)$. The equation of the ellipse is beginning to develop. We can incorporate the center of the ellipse into the equation:

$$\frac{(x-4)^2}{a^2} + \frac{(y-6)^2}{b^2} = 1$$

The distance between the center and one of the vertices is b. The distance between $(4, 9)$ and $(4, 6)$ is 3, so $b = 3$. The distance between the center and one of the foci is c. The distance between $(4, 8)$ and $(4, 6)$ is 2, so $c = 2$. Now that we know b and c, we can find a^2:

$c^2 = b^2 - a^2$

$2^2 = 3^2 - a^2$

$4 = 9 - a^2$

$a^2 = 5$

The equation of the ellipse is $\dfrac{(x-4)^2}{5} + \dfrac{(y-6)^2}{9} = 1$.

Ellipses have a reflection property similar to what we saw with parabolas: If an energy source is placed at one focus of a reflecting surface with elliptical cross-sections, then the energy will be reflected off of the surface and directed to the other focus. The energy source can be light or sound. A whispering gallery is a room designed with elliptical ceilings. A whisper spoken at one focus can be heard clearly across the room at the other focus. St. Paul's Cathedral in London and the Statuary Hall in the United States' Capitol Building are two examples of well-known whispering galleries.

There are other applications of ellipses. The orbits of the planets around the sun are elliptical, with the sun positioned at one of the foci. A tank truck is often in the shape of an ellipse. If its shape were more circular, it would be "top-heavy" and less stable. In addition, any cylinder that is sliced at an angle will yield an elliptical cross-section.

Lesson 9-2 Review

1. Sketch the graph of the ellipse given by the formula $\dfrac{x^2}{100}+\dfrac{y^2}{64}=1$.

    Find the vertices, the foci, and the eccentricity of the ellipse.

2. The vertices of an ellipse are $(\pm5, 0)$ and the foci are $(\pm2, 0)$. Find the equation of the ellipse and its eccentricity. Sketch a graph of the ellipse.

3. The vertices of an ellipse are $(\pm12, 0)$, and the eccentricity of the

    ellipse is $e=\frac{2}{3}$. Find the equation of the ellipse and its foci. Sketch

    a graph of the ellipse.

4. Find the equation of the ellipse with one focus at $(10, 3)$ and vertices located at $(-2, 3)$ and $(12, 3)$.

Lesson 9-3: Hyperbolas

The graphs of ellipses and hyperbolas look very different, but their definitions are very similar. With an *ellipse*, the distances from two fixed foci are *added* together. With a *hyperbola*, the distances from two fixed foci are *subtracted*.

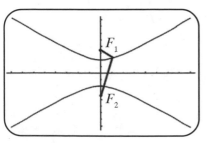

Figure 9.15.

A **hyperbola** is defined to be the set of all points in the plane, the difference of whose distances from two fixed points F_1 and F_2 is a constant. The two fixed points are called the **foci** of the hyperbola. Figure 9.15 shows a hyperbola with foci F_1 and F_2. The line that contains the foci is called the **transverse axis**. The

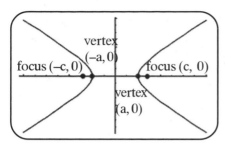

Figure 9.16.

midpoint of the line segment $\overline{F_1F_2}$ is called the **center** of the hyperbola. The line that passes through the center and is perpendicular to the transverse axis is called the **conjugate axis**. A hyperbola consists of two separate curves, or **branches**, that are symmetric with respect to the transverse axis, the conjugate axis, and the center. The points where the hyperbola intersects the transverse axis are called the **vertices** of the hyperbola. Each vertex of a hyperbola lies between the center and one of the foci of the hyperbola.

We will begin by presenting the equation for a hyperbola when the transverse axis is the x-axis. The equation for the hyperbola with center at the origin, foci at $(\pm c, 0)$, and vertices at $(\pm a, 0)$, as shown in Figure 9.16, is:

$$\frac{x^2}{a^2} - \frac{y^2}{b^2} = 1 \text{, where } b^2 = c^2 - a^2$$

If the transverse axis of a hyperbola is the y-axis, then the foci and the vertices of the hyperbola will be located on the y-axis. The equation for the hyperbola with center at the origin, foci at $(0, \pm c)$, and vertices at $(0, \pm a)$, as shown in Figure 9.17, is:

$$\frac{y^2}{a^2} - \frac{x^2}{b^2} = 1 \text{, where } b^2 = c^2 - a^2$$

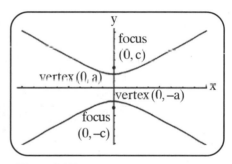

Figure 9.17.

Keep in mind that a hyperbola involves a *difference* of two quadratic terms. The *positive* term determines which axis is the transverse axis. The foci and vertices of a hyperbola lie on the transverse axis.

Example 1

Find the coordinates of the foci and the vertices of the hyperbola described by $\dfrac{x^2}{9} - \dfrac{y^2}{16} = 1$, and sketch its graph.

Solution: Because the positive term involves x, the x-axis is the transverse axis. The foci and the vertices of the hyperbola will lie on the x-axis. The vertices of the hyperbola are $(\pm 3, 0)$. The foci are located at $(\pm c, 0)$, where the x-coordinates of the foci must satisfy the equation $b^2 = c^2 - a^2$:

$$b^2 = c^2 - a^2$$
$$4^2 = c^2 - 3^2$$
$$16 = c^2 - 9$$
$$c^2 = 25$$

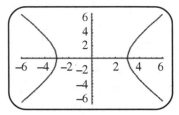

The foci are located at $(\pm 5, 0)$. The graph of the hyperbola is shown in Figure 9.18.

Figure 9.18.

Example 2

Find the equation of the hyperbola that has one vertex at $(0, 2)$ and foci at $(0, \pm 3)$.

Solution: The foci and the vertices of the hyperbola are located on the y-axis, so the transverse axis is the y-axis. The center of the hyperbola is located at the midpoint of the line segment connecting the two foci. The midpoint of the line segment with endpoints $(0, 3)$ and $(0, -3)$ is the origin. With this hyperbola, $a^2 = 4$ and $c^2 = 9$. From this, we can find b^2:

$$b^2 = c^2 - a^2$$
$$b^2 = 9 - 4$$
$$b^2 = 5$$

The equation of the hyperbola is $\dfrac{y^2}{4} - \dfrac{x^2}{5} = 1$.

We will now turn our attention to the asymptotic behavior of hyperbolas. We have analyzed the asymptotic behavior of functions in previous chapters. Remember that four of the six trigonometric functions have vertical asymptotes, and some of the inverse trigonometric functions have horizontal asymptotes. Hyperbolas have *oblique* asymptotes.

The *asymptotic* behavior of a function describes how the function behaves when either the dependent variable or the independent variable become very large in magnitude, or tend towards ∞. The asymptotic behavior of a hyperbola is very important in describing the graph of the hyperbola.

For large values of x (and corresponding large values of y), a hyperbola behaves as a linear function. To describe this linear behavior, start with the equation of a hyperbola (the particular form does not matter) and simplify it under the assumption that x and y are much, much larger than 1. I will examine hyperbolas of the form $\frac{y^2}{a^2} - \frac{x^2}{b^2} = 1$. Isolate the term involving y by adding $\frac{x^2}{b^2}$ to both sides: $\frac{y^2}{a^2} = \frac{x^2}{b^2} + 1$.

If we assume that x is very, very large, so that adding 1 to the term $\frac{x^2}{b^2}$ is hardly noticeable, we see that $\frac{y^2}{a^2} \sim \frac{x^2}{b^2}$. I am using the symbol ~ to mean "behaves like." I cannot write that $\frac{y^2}{a^2}$ and $\frac{x^2}{b^2}$ are equal, because they are not. They are, however, very close. If $\frac{y^2}{a^2} \sim \frac{x^2}{b^2}$, then $y^2 \sim \frac{a^2 x^2}{b^2}$, and taking the square root of both sides gives $y \sim \pm \frac{ax}{b}$. The two lines, $y = \frac{ax}{b}$ and $y = -\frac{ax}{b}$, are called the **oblique asymptotes** of the hyperbola $\frac{y^2}{a^2} - \frac{x^2}{b^2} = 1$. Along the same reasoning, the oblique asymptotes of the hyperbola $\frac{x^2}{a^2} - \frac{y^2}{b^2} = 1$ are $y = \pm \frac{bx}{a}$.

These equations are not worth memorizing. If you understand the process of finding the asymptotes, you will avoid any confusion as to whether the slope of the oblique asymptotes are $\pm \frac{a}{b}$ or $\pm \frac{b}{a}$. Personally, I have difficulty memorizing formulas and prefer to understand where the formulas come from. It may require an extra step or two when solving a problem, but those extra steps reinforce the concepts behind finding asymptotes. The key to finding the oblique asymptotes of a hyperbola is to ignore the constant in the standard equation for the hyperbola, and only focus on the terms involving x and y.

Example 3

Find the asymptotes of the following hyperbolas:

a. $\dfrac{y^2}{36} - \dfrac{x^2}{25} = 1$

b. $\dfrac{x^2}{121} - \dfrac{y^2}{144} = 1$

Solution: Isolate the term that involves y and then assume that the constant 1 is negligible:

a. $\dfrac{y^2}{36} - \dfrac{x^2}{25} = 1:$ $\dfrac{y^2}{36} = \dfrac{x^2}{25} + 1$

$$\dfrac{y^2}{36} \sim \dfrac{x^2}{25}$$

$$y^2 \sim \dfrac{36x^2}{25}$$

$$y \sim \pm \dfrac{6x}{5}$$

The oblique asymptotes are: $y = \pm \dfrac{6x}{5}$.

b. $\dfrac{x^2}{121} - \dfrac{y^2}{144} = 1:$ $\dfrac{x^2}{121} - \dfrac{y^2}{144} = 1$

$$\dfrac{y^2}{144} = \dfrac{x^2}{121} - 1$$

$$\dfrac{y^2}{144} \sim \dfrac{x^2}{121}$$

$$y^2 \sim \dfrac{144x^2}{121}$$

$$y \sim \pm \dfrac{12x}{11}$$

The oblique asymptotes are: $y = \pm \dfrac{12x}{11}$.

The graph of a hyperbola can be translated horizontally or vertically. When a hyperbola is translated, every point on the hyperbola is translated by the same amount. In particular, the center, the foci, and the vertices of the hyperbola are translated. The equations $\dfrac{x^2}{a^2} - \dfrac{y^2}{b^2} = 1$ and $\dfrac{y^2}{a^2} - \dfrac{x^2}{b^2} = 1$ describe hyperbolas whose center is (0, 0). We could write these equations

as $\dfrac{(x-0)^2}{a^2}+\dfrac{(y-0)^2}{b^2}=1$ and $\dfrac{(y-0)^2}{a^2}-\dfrac{(x-0)^2}{b^2}=1$, if we wanted to explicitly include the location of the center of the hyperbola. In general, the equa-
tions of hyperbolas with centers located at (h, k) can be written as:

$$\dfrac{(x-h)^2}{a^2}-\dfrac{(y-k)^2}{b^2}=1 \text{ and } \dfrac{(y-k)^2}{a^2}-\dfrac{(x-h)^2}{b^2}=1$$

Be sure to associate the x-coordinate of the center with the term that
involves x in the equation of the hyperbola, and the y-coordinate of the
center with the term that involves y in the equation of the hyperbola. We
can find c by finding the distance between the center and one focus. The
distance between the center and one of the vertices is a. The asymptotes
will be translated by the same amount that the center is translated. The
following table may help you organize this information and keep it all
straight.

Equation for the hyperbola	$\dfrac{(x-h)^2}{a^2}-\dfrac{(y-k)^2}{b^2}=1$	$\dfrac{(y-k)^2}{a^2}-\dfrac{(x-h)^2}{b^2}=1$
Transverse axis	Parallel to the x-axis	Parallel to the y-axis
Vertices	$(h \pm a, k)$	$(h, k \pm a)$
Foci	$(h \pm c, k)$	$(h, k \pm c)$
Asymptotes	$y-k=\pm\dfrac{b}{a}(x-h)$	$y-k=\pm\dfrac{a}{b}(x-h)$

Example 4

Find the equation of a hyperbola with center $(-3, 1)$, one focus at
$(-3, 6)$, and one vertex at $(-3, 4)$.

Solution: The center, the focus and the vertex all lie on the vertical line $x = -3$, so the transverse axis is parallel to the y-axis. We
can therefore start with the equation $\dfrac{(y-k)^2}{a^2}-\dfrac{(x-h)^2}{b^2}=1$ and start
filling in the missing pieces. The center of the hyperbola is $(-3, 1)$,

so $h = -3$ and $k = 1$. Our equation becomes $\dfrac{(y-1)^2}{a^2} - \dfrac{(x+3)^2}{b^2} = 1$. To find c, we need to find the distance between the center and the focus: this distance is 5, so $c = 5$. To find a, we need to find the distance between the center and one vertex: the distance between $(-3, 1)$ and $(-3, 4)$ is $a = 3$. We can use the equation $b^2 = c^2 - a^2$ to find b^2:

$$b^2 = c^2 - a^2 = 5^2 - 3^2 = 25 - 9 = 16$$

The equation of the hyperbola is:

$$\frac{(y-1)^2}{9} - \frac{(x+3)^2}{16} = 1$$

In general, the equation of the oblique asymptotes is: $y - k = \pm\dfrac{a}{b}(x - h)$. Substituting in for the values of a, b, h, and k gives:

$$y - 1 = \pm\frac{3}{4}(x + 3)$$

Example 5

Find the vertices, the foci, and the oblique asymptotes of the hyperbola $x^2 - 3y^2 + 8x - 6y + 4 = 0$.

Solution: We need to complete the square in both x and y and put the resulting equation in standard form for a hyperbola:

$$x^2 - 3y^2 + 8x - 6y + 4 = 0$$

Group the terms together $(x^2 + 8x) - (3y^2 + 6y) = -4$

Factor 3 from both terms involving y

$$(x^2 + 8x) - 3(y^2 + 2y) = -4$$

Complete the square $(x^2 + 8x + 16 - 16) - 3(y^2 + 2y + 1 - 1) = -4$

Pull the unnecessary constants out of the parentheses
(Make sure you multiply the constant by the coefficient in front.)

$$(x^2 + 8x + 16) - 16 - 3(y^2 + 2y + 1) + 3 = -4$$

Factor the perfect squares $(x + 4)^2 - 16 - 3(y + 1)^2 + 3 = -4$

Move the constants to the other side of the equation

$$(x+4)^2 - 3(y+1)^2 = 9$$

Divide by 9 so that the constant term is 1

$$\frac{(x+4)^2}{9} - \frac{3(y+1)^2}{9} = 1$$

Cancel the 3's from the term involving y

$$\frac{(x+4)^2}{9} - \frac{(y+1)^2}{3} = 1$$

Now that the hyperbola is in standard form, we can read off the important information. The center is $(-4, -1)$, $a^2 = 9$, and $b^2 = 3$. We can solve for c:

$$b^2 = c^2 - a^2$$
$$3 = c^2 - 9$$
$$c^2 = 12$$
$$c = \sqrt{12}$$

The transverse axis is parallel to the x-axis, the vertices are found using the expression $(h \pm a, k)$:

$$(h \pm a, k) = (-4 \pm 3, -1)$$

The vertices are $(-7, -1)$ and $(-1, -1)$. The foci are found using the expression $(h \pm c, k)$:

$$(h \pm c, k) = \left(-4 \pm \sqrt{12}, -1\right)$$

The foci are $\left(-4 + \sqrt{12}, -1\right)$ and $\left(-4 - \sqrt{12}, -1\right)$.

The oblique asymptotes are found using the equation

$$y - k = \pm\frac{b}{a}(x - h): \qquad y - k = \pm\frac{b}{a}(x - h)$$

$$y + 1 = \pm\frac{\sqrt{3}}{3}(x + 4)$$

The oblique asymptotes are: $y = -1 \pm \dfrac{\sqrt{3}}{3}(x + 4)$.

There are many applications of hyperbolas in science and engineering. The properties of a hyperbola are used in radar tracking stations and navigation. An object can be located by sending sound waves from two sources. The sound waves from each source travel in concentric circles, and the intersection of the concentric circles from the two sources is hyperbolic. Hyperbolas are also used in astronomy. Although we typically think of the orbit of an object as an ellipse, the orbit of a comet is a hyperbola. Hyperbolas also appear when two quantities are inversely related to each other. For example, the volume and the pressure of a gas held at a fixed temperature are inversely related to each other, and their relationship is hyperbolic in nature.

Lesson 9-3 Review

1. Find the coordinates of the foci and the vertices of the hyperbola described by $\frac{x^2}{25} - \frac{y^2}{49} = 1$. Find the equation of the oblique asymptotes and sketch its graph.

2. Find the equation of the hyperbola that has one vertex at $(3, 0)$ and foci at $(\pm 5, 0)$.

3. Find the equation of a hyperbola with center $(1, 4)$, one focus at $(-2, 4)$, and one vertex at $(0, 4)$.

4. Find the vertices, the foci, and the oblique asymptotes of the hyperbola $-x^2 + 2y^2 + 2x + 8y + 3 = 0$.

Lesson 9-4: Rotation of Axes

Though the definitions of the conics were very different, the equations for parabolas, ellipses, and hyperbolas shared some overlapping characteristics. They all involved at least one variable squared. A parabola only had one variable squared; the highest power of the other variable was 1. An ellipse had two terms that involved variables that were squared, and these higher-powered terms had the same sign. A hyperbola also had two terms that involved variables that were squared, but these higher-powered terms had opposite signs.

In general, the equation of a conic whose center is the origin can be written as $Ax^2 + Bx + Cy^2 + Dy + E = 0$. For a parabola, either A and D, or B and C, are equal to 0. For a circle, B and D are equal to 0 and $A = C$. For an ellipse, B and D are equal to 0 and A and C have the same sign. For a hyperbola, B and D are equal to 0 and A and C have the opposite sign.

The equation $Ax^2 + Bx + Cy^2 + Dy + E = 0$ can also describe the graph of a conic whose center is not the origin. If a conic is shifted horizontally, then $B \neq 0$, and if the conic is shifted vertically, $D \neq 0$. Notice that in the equation $Ax^2 + Bx + Cy^2 + Dy + E = 0$, there is no mixed term, or term that involves the product of x and y. This mixed term appears when the conic is rotated in the coordinate plane.

The equation $Ax^2 + Bxy + Cy^2 + Dx + Ey + F = 0$ represents the most general form of a conic. Besides parabolas, ellipses, and hyperbolas, the equation $Ax^2 + Bxy + Cy^2 + Dx + Ey + F = 0$ can be used to describe a single point, a line, a pair of lines, or the empty set (meaning that there are no values of x and y for which the equation will be satisfied). Parabolas, ellipses, and hyperbolas are called **non-degenerate conics**. The other cases are called **degenerate conics** and are not particularly interesting. The non-degenerate conic associated with a particular equation can be identified from the constants A, B, and C. If $B^2 - 4AC = 0$, then the conic will be a parabola. If $B^2 - 4AC < 0$, then the conic will be an ellipse. If $B^2 - 4AC > 0$, then the conic will be a hyperbola. The expression $B^2 - 4AC$ is called the **discriminant** of the equation.

Example 1

Identify the conic $2x^2 - 4xy + 2y^2 - 5x - 5 = 0$.

Solution: Evaluate the discriminant of the equation:

$B^2 - 4AC = (-4)^2 - 4(2)(2) = 16 - 16 = 0$

The conic is a parabola.

To understand the graph of a particular conic equation, we will need to rotate the coordinate axes through a particular angle and eliminate the term that involves the product of x and y. Rotating the coordinate axes is fairly straightforward if we use a polar coordinate grid, and trigonometry will be instrumental in making this transformation.

We will rotate our axis by an angle α about the origin, as shown in Figure 9.19. We will need to describe points using both the old coordinate system and the new coordinate system.

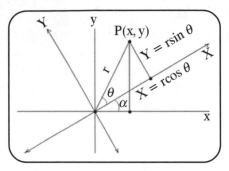

Figure 9.19.

Suppose a point P has coordinates (x, y) in the old coordinate system. We can represent P using polar coordinates: let r represent the distance between P and the origin, and let θ represent the angle that the segment $\overline{OP}$ makes with the new positive x-axis. Then $x = r \cos(\theta + \alpha)$ and $y = r \sin(\theta + \alpha)$. The coordinates of the point relative to the new coordinate system, denoted (X, Y), can be found using the equations $X = r \cos\theta$ and $Y = r \sin\theta$. Our goal is to find equations for x and y in terms of X, Y, and α. We can use the sum formulas for the sine and cosine functions:

$$x = r \cos(\theta + \alpha) \qquad\qquad y = r \sin(\theta + \alpha)$$
$$x = r(\cos\theta \cos \alpha - \sin\theta \sin \alpha) \quad y = r(\sin\theta \cos \alpha + \cos\theta \sin \alpha)$$
$$x = r \cos\theta \cos \alpha - r \sin\theta \sin \alpha \quad y = r \sin\theta \cos \alpha + r \cos\theta \sin \alpha$$
$$x = X \cos \alpha - Y \sin \alpha \qquad\qquad y = Y \cos \alpha + X \sin \alpha$$

We can also solve for X and Y in terms of x and y:

$X = x \cos \alpha + y \sin \alpha$ and $Y = y \cos \alpha - x \sin \alpha$

These two sets of equations are related: changing from XY-coordinates to xy-coordinates is equivalent to rotating by an angle $-\alpha$.

Example 2

If the coordinate axes are rotated through an angle of 45°, find the following:

a. The XY-coordinates of the point with xy-coordinates $(2, 4)$

b. The xy-coordinates of the point with XY-coordinates $(-1, 1)$

Solution:

a. Use the formulas $X = x \cos \alpha + y \sin \alpha$ and $Y = y \cos \alpha - x \sin \alpha$ to find X and Y:

$$X = x \cos \alpha + y \sin \alpha \qquad Y = y \cos \alpha - x \sin \alpha$$
$$X = 2 \cos 45° + 4 \sin 45° \qquad Y = 4 \cos 45° - 2 \sin 45°$$
$$X = 2 \cdot \frac{\sqrt{2}}{2} + 4 \cdot \frac{\sqrt{2}}{2} \qquad Y = 4 \cdot \frac{\sqrt{2}}{2} - 2 \cdot \frac{\sqrt{2}}{2}$$
$$X = 3\sqrt{2} \qquad\qquad Y = \sqrt{2}$$

The coordinates of $(2, 4)$ are $\left(3\sqrt{2}, \sqrt{2}\right)$. Figure 9.20 will help you visualize this change in coordinates.

b. Use the formulas $x = X \cos \alpha - Y \sin \alpha$ and $y = Y \cos \alpha + X \sin \alpha$ to find x and y:

$$x = X \cos \alpha - Y \sin \alpha \qquad y = Y \cos \alpha + X \sin \alpha$$
$$x = -1 \cos 45° - 1 \sin 45° \qquad y = 1 \cos 45° + (-1) \sin 45°$$
$$x = -1 \cdot \frac{\sqrt{2}}{2} - 1 \cdot \frac{\sqrt{2}}{2} \qquad y = 1 \cdot \frac{\sqrt{2}}{2} - 1 \cdot \frac{\sqrt{2}}{2}$$
$$X = -\sqrt{2} \qquad\qquad Y = 0$$

The coordinates of $(-1, 1)$ are $\left(-\sqrt{2}, 0\right)$. Figure 9.21 will help you visualize this change in coordinates.

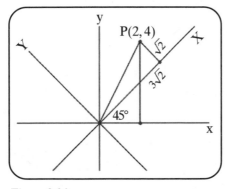

Figure 9.20.

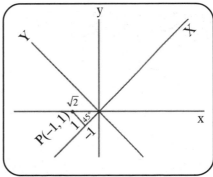

Figure 9.21.

As I mentioned earlier, the equation $Ax^2 + Bxy + Cy^2 + Dx + Ey + F = 0$ represents the most general form of a conic. To understand the graph of a particular conic equation, we will need to rotate the coordinate axes through a particular angle and eliminate the term that involves the product of x and y. The angle of rotation depends on the constants A, B, and C, and is the acute angle α that satisfies the equation:

$$\tan 2\alpha = \frac{B}{A-C}$$

Rotating the coordinate axis by an angle α will remove the xy term in the conic equation $Ax^2 + Bxy + Cy^2 + Dx + Ey + F = 0$. To find the angle α, follow the following procedure:

1. Find $\tan 2\alpha$.

2. Using the Pythagorean Theorem, find $\cos 2\alpha$.

3. Find $\sin \alpha$ and $\cos \alpha$ using the equations $\sin\alpha = \sqrt{\dfrac{1-\cos 2\alpha}{2}}$ and $\cos\alpha = \sqrt{\dfrac{1+\cos 2\alpha}{2}}$. We will take the positive square roots because α will be in Quadrant I.

4. Substitute the values for $\sin \alpha$ and $\cos \alpha$ into the equations $x = X \cos \alpha - Y \sin \alpha$ and $y = Y \cos \alpha + X \sin \alpha$.

5. Substitute the equations $x = X \cos \alpha - Y \sin \alpha$ and $y = Y \cos \alpha + X \sin \alpha$ into the conic equation and simplify. The xy term should subtract out after this simplification.

Example 3

Find the angle α that the coordinate axes should be rotated to remove the xy term in the equation $-7x^2 + 48xy + 7y^2 - 25 = 0$, and sketch its graph.

Solution: Follow the procedure outlined earlier.

1. In this situation, $A = -7$, $B = 48$, and $C = 7$, and α must be chosen so that $\tan 2\alpha = \dfrac{B}{A-C}$:

$$\tan 2\alpha = \frac{B}{A-C} = \frac{48}{-7-7} = -\frac{48}{14} = -\frac{24}{7}$$

From the sign of $\tan 2\alpha = \dfrac{B}{A-C}$, we see that 2α is in Quadrant II.

2. Using the Pythagorean Theorem, we see that $\cos 2\alpha = -\frac{7}{25}$.

3. Find $\sin \alpha$ and $\cos \alpha$:

$$\sin \alpha = \sqrt{\frac{1-\cos 2\alpha}{2}} = \sqrt{\frac{1-\left(-\frac{7}{25}\right)}{2}} = \sqrt{\frac{\frac{32}{25}}{2}} = \sqrt{\frac{16}{25}} = \frac{4}{5}$$

$$\cos \alpha = \sqrt{\frac{1+\cos 2\alpha}{2}} = \sqrt{\frac{1+\left(-\frac{7}{25}\right)}{2}} = \sqrt{\frac{\frac{18}{25}}{2}} = \sqrt{\frac{9}{25}} = \frac{3}{5}$$

4. Substitute the values for $\sin \alpha$ and $\cos \alpha$ into the equations $x = X\cos \alpha - Y \sin \alpha$ and $y = Y\cos \alpha + X \sin \alpha$:

$$x = \tfrac{3}{5}X - \tfrac{4}{5}Y$$

$$y = \tfrac{3}{5}Y + \tfrac{4}{5}X$$

5. Substitute the equations $x = \tfrac{3}{5}X - \tfrac{4}{5}Y$ and $y = \tfrac{3}{5}Y + \tfrac{4}{5}X$ into the conic equation and simplify:

$$-7x^2 + 48xy + 7y^2 - 25 = 0$$

$$-7\left(\tfrac{3}{5}X - \tfrac{4}{5}Y\right)^2 + 48\left(\tfrac{3}{5}X - \tfrac{4}{5}Y\right)\left(\tfrac{3}{5}Y + \tfrac{4}{5}X\right) + 7\left(\tfrac{3}{5}Y + \tfrac{4}{5}X\right)^2 - 25 = 0$$

$$-7\left(\tfrac{9}{25}X^2 - \tfrac{24}{25}XY + \tfrac{16}{25}Y^2\right) + 48\left(\tfrac{12}{25}X^2 - \tfrac{7}{25}XY - \tfrac{12}{25}Y^2\right)$$
$$+ 7\left(\tfrac{9}{25}Y^2 + \tfrac{24}{25}XY + \tfrac{16}{25}X\right) - 25 = 0$$

$$-\tfrac{63}{25}X^2 + \tfrac{168}{25}XY - \tfrac{112}{25}Y^2 + \tfrac{576}{25}X^2 - \tfrac{336}{25}XY - \tfrac{576}{25}Y^2$$
$$+ \tfrac{63}{25}Y^2 + \tfrac{168}{25}XY + \tfrac{112}{25}X^2 - 25 = 0$$

$$\tfrac{625}{25}X^2 - \tfrac{625}{25}Y^2 - 25 = 0$$

$$\tfrac{625}{25}X^2 - \tfrac{625}{25}Y^2 = 25$$

$$625X^2 - 625Y^2 = 625$$

$$X^2 - Y^2 = 1$$

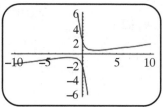

Figure 9.22.

The angle of rotation is $\alpha = \cos^{-1}\tfrac{3}{5}$, and the graph of the hyperbola is shown in Figure 9.22.

Lesson 9-4 Review

1. Identify the following conics:

    a. $x^2 + 3xy - 2y^2 + 3x + 2y - 5 = 0$

    b. $9x^2 + 12xy + 4y^2 - x - y = 0$

    c. $10x^2 - 12xy + 4y^2 - x - y - 10 = 0$

2. Find the angle α that the coordinate axes should be rotated to remove the xy term in the equation $41x^2 + 24xy + 34y^2 - 25 = 0$, and sketch its graph.

Lesson 9-5: Conics in Polar Coordinates

Although the graphs of the conics appear to be very different, there are several unifying characteristics. The definition of a parabola involved the relationship between a point, the focus, and a line, the directrix. The earlier definitions of an ellipse and a hyperbola involved combining distances between two foci. We discussed the concept of eccentricity for an ellipse, but not for a parabola or a hyperbola. We can actually define all of the conics in terms of a point (the focus) and a line (the directrix). We can expand the idea of eccentricity to apply to all of the conics as well, and the eccentricity of a conic can be used to distinguish between the different conics. This alternative method for developing the three conics is based on their polar representations.

Let F be a fixed point, called the **focus**, l a fixed line, called the **directrix**, and e a fixed positive number, called the **eccentricity**. Then the set of all points P such that the ratio of the distance from P to F to the distance from P to 1, is equal to the constant e is a conic. In other words, if $d(P, F)$ denotes the distance from P to F, and $d(P, 1)$ represents the distance from P to 1, then the set of all points

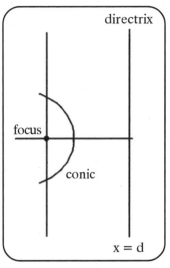

directrix

focus

conic

$x = d$

Figure 9.23.

P such that $\dfrac{d(P,F)}{d(P,\ell)} = e$ is a conic. The conic is

a parabola if $e = 1$, an ellipse if $e < 1$, or a hyperbola if $e > 1$.

Geometrically, suppose that the focus is located at the origin and the directrix is the vertical line $x = d$. If P is a point on the conic with rectangular coordinates (x, y), then its location in polar coordinates is $(r \cos \theta, r \sin \theta)$, where $x^2 + y^2 = r^2$ and θ is the angle formed by the ray $\overrightarrow{OP}$ and the positive x-axis, as shown in Figure 9.23.

The distance from P to the focus is r, and the distance from P to the line is $d - r \cos \theta$. If the ratio of these distances satisfies the

equation $\dfrac{d(P,F)}{d(P,\ell)} = e$, then $r = e(d - r \cos \theta)$. If we square both sides of this

equation and convert to rectangular coordinates, the result will be one of the conic equations we studied earlier in the chapter.

In general, a polar equation of the form $r = \dfrac{ed}{1 \pm e \cos \theta}$ or $r = \dfrac{ed}{1 \pm e \sin \theta}$
represents a conic with one focus at the origin with eccentricity e. The conic is a parabola if $e = 1$, an ellipse if $e < 1$, or a hyperbola if $e > 1$. If the conic is an ellipse, then the *major* axis is a line that passes through the focus that is perpendicular to the directrix. If the conic is a hyperbola, the *transverse* axis is a line that passes through the focus that is perpendicular to the directrix. The eccentricity e satisfies the equation $e = \dfrac{c}{a}$,
where c is the distance from the center to the focus, and a is the distance from the center to a vertex. The following table summarizes the properties of each of these forms of a conic.

$r = \dfrac{ed}{1+e\cos\theta}$	Directrix is perpendicular to the x-axis and is a distance d units to the right of the origin.
$r = \dfrac{ed}{1-e\cos\theta}$	Directrix is perpendicular to the x-axis and is a distance d units to the left of the origin.
$r = \dfrac{ed}{1+e\sin\theta}$	Directrix is perpendicular to the y-axis and is a distance d units above the origin.
$r = \dfrac{ed}{1-e\sin\theta}$	Directrix is perpendicular to the y-axis and is a distance d units below the origin.

Example 1

Describe and sketch the graph of the equation $r = \dfrac{2}{1+3\cos\theta}$.

Solution: This equation is of the form $r = \dfrac{ed}{1+e\cos\theta}$, with $e = 3$ and $e \cdot d = 2$, or $d = \frac{2}{3}$. Because $e > 1$, this is the equation of a hyperbola with one focus at the origin. The directrix is the line $x = \frac{2}{3}$. The transverse axis is along the x-axis. The vertices can be found by evaluating this function when $\theta = 0$ and $\theta = \pi$. The *polar* coordinates of the vertices are the points $\left(\frac{1}{2},0\right)$ and $(-1, \pi)$. The *rectangular* coordinates of the vertices are the points $\left(\frac{1}{2},0\right)$ and $(1, 0)$. The center of the hyperbola is the midpoint of the line segment connecting $\left(\frac{1}{2},0\right)$ and $(1, 0)$, which is the point $\left(\frac{3}{4},0\right)$. The distance between the center and one of the vertices is a: $a = \frac{1}{4}$.

Using the fact that $e = 3$ and $a = \frac{1}{4}$, we can find c: $e = \dfrac{c}{a}$

$$3 = \dfrac{c}{\frac{1}{4}}$$

$$c = \dfrac{3}{4}$$

Now that we know a and c, we can find b: $b^2 = c^2 - a^2$

$$b^2 = \left(\tfrac{3}{4}\right)^2 - \left(\tfrac{1}{4}\right)^2$$

$$b^2 = \tfrac{9}{16} - \tfrac{1}{16}$$

$$b^2 = \tfrac{1}{2}$$

$$b = \tfrac{\sqrt{2}}{2}$$

The equations for the oblique asymptotes are $y - k = \pm \tfrac{b}{a}(x - h)$.

The center is $\left(\tfrac{3}{4}, 0\right)$, which means that

$h = \tfrac{3}{4}$ and $k = 0$:

$$y = \pm \frac{\tfrac{\sqrt{2}}{2}}{\tfrac{1}{4}}\left(x - \tfrac{3}{4}\right)$$

$$y = \pm 2\sqrt{2}\left(x - \tfrac{3}{4}\right)$$

The equivalent equation for the hyperbola in rectangular coordinates is

$16\left(x - \tfrac{3}{4}\right)^2 - 2y^2 = 1$, and its graph is shown in Figure 9.24.

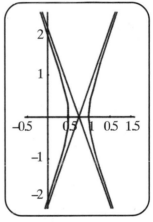

Figure 9.24.

You have been introduced to quite a few formulas in this chapter, and throughout this book. It is important to understand how the formulas were developed. Trigonometry plays an important role in analytic geometry, especially when analyzing rotational transformations. Being able to transform an equation from one set of coordinates to another set will enable you to make use of the different strengths of various coordinate systems. For example, the polar form of a conic is much more compact to write, but may be less intuitive to analyze and visualize.

Polar coordinates and rectangular coordinates are not the only two possible coordinate systems, but they are two of the most popular. Different coordinate systems give different perspectives on a problem. Understanding the process of changing from rectangular coordinates to polar coordinates, or changing from one set of rectangular coordinates to another set of rectangular coordinates through rotation or translation can be a valuable skill, especially when working out physics and engineering problems where changing coordinate systems is equivalent to changing reference points. Changing coordinate systems can make an unsolvable problem suddenly solvable.

Lesson 9-5 Review

Discuss and sketch the graphs of the following conics:

1. $r = \dfrac{4}{4 - 2\cos\theta}$

2. $r = \dfrac{2}{1 + \sin\theta}$

Answer Key
Lesson 9-1 Review

1. a. $y^2 = -10x$: $a = -10$,

   focus $\left(-\frac{10}{4}, 0\right)$,

   directrix: $x = \frac{10}{4}$,

   axis of symmetry is the x-axis,

   and the parabola opens upward.

   b. $x^2 = 4y$: $a = 4$,

   focus $(0, 1)$,

   directrix: $y = -1$,

   axis of symmetry is the y-axis,

   and the parabola opens to the left.

2. $x^2 = \frac{16}{3}y$

3. Axis of symmetry: $x = 3$,

   directrix: $y = -8$, $(x - 3)^2 = 24(y + 2)$

4. Complete the square to find the equation for the standard form of the parabola: $(x + 3)^2 = -1(x - 9)$.

   $a = -1$,

   vertex: $(9, -3)$,

   focus: $\left(\frac{35}{4}, -3\right)$,

   directrix: $x = \frac{37}{4}$.

5. The receiver should be placed 0.64 feet from the base of the dish, along its axis of symmetry.

Lesson 9-2 Review

1. The vertices are $(\pm 10, 0)$,
   the foci are $(\pm 6, 0)$,

   and the eccentricity is $e = \dfrac{c}{a} = \dfrac{6}{10} = \dfrac{3}{5}$.

   The graph of $\dfrac{x^2}{100} + \dfrac{y^2}{64} = 1$ is shown in
   Figure 9.25.

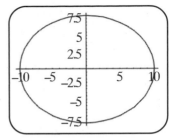

Figure 9.25.

2. The equation of the ellipse is $\dfrac{x^2}{25} + \dfrac{y^2}{21} = 1$,

   and its eccentricity is $e = \dfrac{c}{a} = \dfrac{2}{5}$.

   The graph of $\dfrac{x^2}{25} + \dfrac{y^2}{21} = 1$ is shown in
   Figure 9.26.

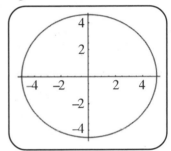

Figure 9.26.

3. The equation of the ellipse is $\dfrac{x^2}{144} + \dfrac{y^2}{80} = 1$,

   and the foci are located at $(\pm 8, 0)$.

   The graph of $\dfrac{x^2}{144} + \dfrac{y^2}{80} = 1$ is shown in
   Figure 9.27.

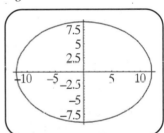

Figure 9.27.

4. Major axis: $y = 3$,
   center: $(5, 3)$,
   the equation of the ellipse is

   $$\dfrac{(x-5)^2}{49} + \dfrac{(y-3)^2}{24} = 1.$$

Lesson 9-3 Review

1. Foci: $\left(\pm\sqrt{74}, 0\right)$,

   vertices: $(\pm 5, 0)$,

   oblique asymptotes: $y = \pm\dfrac{7x}{5}$.

   The graph of $\dfrac{x^2}{25} - \dfrac{y^2}{49} = 1$ is shown in
   Figure 9.28.

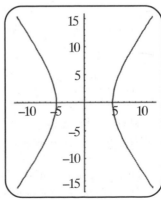

Figure 9.28.

2. $\dfrac{x^2}{9} - \dfrac{y^2}{16} = 1$

3. $\dfrac{(x-1)^2}{1} - \dfrac{(y-4)^2}{8} = 1$

4. After completing the square, we have $\dfrac{(y+2)^2}{2} - \dfrac{(x-1)^2}{4} = 1$.

   The center: $(1,-2)$, $a = \sqrt{2}$, $b = 2$, $c = \sqrt{6}$.

   The vertices: $\left(1, -2 \pm \sqrt{2}\right)$, foci: $\left(1, -2 \pm \sqrt{6}\right)$,

   The oblique asymptotes: $y + 2 = \pm \frac{\sqrt{2}}{2}(x-1)$

Lesson 9-4 Review

1. a. $x^2 + 3xy - 2y^2 + 3x + 2y - 5 = 0$: $B^2 - 4AC = 17 > 0$: hyperbola

   b. $9x^2 + 12xy + 4y^2 - x - y = 0$: $B^2 - 4AC = 0$: parabola

   c. $10x^2 - 12xy + 4y^2 - x - y - 10 = 0$: $B^2 - 4AC = -16 < 0$: ellipse

2. $\tan 2\alpha = \frac{24}{7}$,

   $\cos 2\alpha = \frac{7}{25}$,

   $\cos \alpha = \sqrt{\dfrac{1 + \frac{7}{25}}{2}} = \dfrac{4}{5}$,

   $\alpha = \cos^{-1} \frac{4}{5} \approx 36.9°$.

   The graph of $41x^2 + 24xy + 34y^2 - 25 = 0$ is show in Figure 9.29.

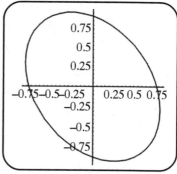

Figure 9.29.

Lesson 9-5 Review

1. $r = \dfrac{4}{4 - 2\cos\theta} = \dfrac{1}{1 - \frac{1}{2}\cos\theta}$: eccentricity $e = \frac{1}{2} < 1$.

   The conic is an ellipse.

   The directrix is perpendicular to the x-axis and is 2 units to the left of the origin: $x = -2$.

   One focus is the origin. The major axis is the x-axis.

   To find the vertices, evaluate the function at $\theta = 0$ and $\theta = \pi$.

The polar coordinates of the vertices are $(2,0)$ and $\left(\frac{2}{3},\pi\right)$.

The equivalent rectangular coordinates are $(2,0)$ and $\left(-\frac{2}{3},0\right)$.

The center is $\left(\frac{2}{3},0\right)$.

The distance from the vertex to the center is $a = \frac{4}{3}$.

The eccentricity is $e = \frac{1}{2} = \frac{c}{a}$, so $c = \frac{2}{3}$.

The foci are at $(0,0)$ and $\left(\frac{4}{3},0\right)$.

$b^2 = a^2 - c^2 = \frac{12}{9} = \frac{4}{3}$.

The equation of the ellipse is

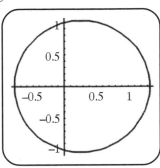

$\dfrac{9\left(x-\frac{2}{3}\right)^2}{16} + \dfrac{4y^2}{3} = 1$, and its graph is shown in Figure 9.30.

Figure 9.30.

2. $r = \dfrac{2}{1+\sin\theta}$: eccentricity $e = 1$.

The conic is a parabola.

The directrix is perpendicular to the y-axis, and is 2 units above the origin: $y = 2$. The focus is the origin.
The axis of symmetry is the x-axis.

The vertex can be found by evaluating the function at $\theta = \frac{\pi}{2}$.

The polar coordinates of the vertex are $\left(1,\frac{\pi}{2}\right)$.

The equivalent rectangular coordinates are $(0,1)$.

The parabola opens downward, and passes through the points $(\pm 2, 0)$.

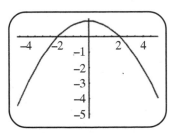

The parabola is of the form $x^2 = a(y-1)$, and $a = -4$, so the equation of the parabola is $x^2 = -4(y-1)$, and its graph is shown in Figure 9.31.

Figure 9.31.

Index

About the Author

DENISE SZECSEI earned Bachelor of Science degrees in physics, chemistry, and mathematics from the University of Redlands, and she was greatly influenced by the educational environment cultivated through the Johnston Center for Integrative Studies. After graduating from the University of Redlands, she served as a technical instructor in the U.S. Navy. After completing her military service, she earned a Ph.D in mathematics from the Florida State University. She has been teaching since 1985.

The Essential Help You Need When Your Textbooks Just Aren't Making the Grade!